The

Cocaine

Kids

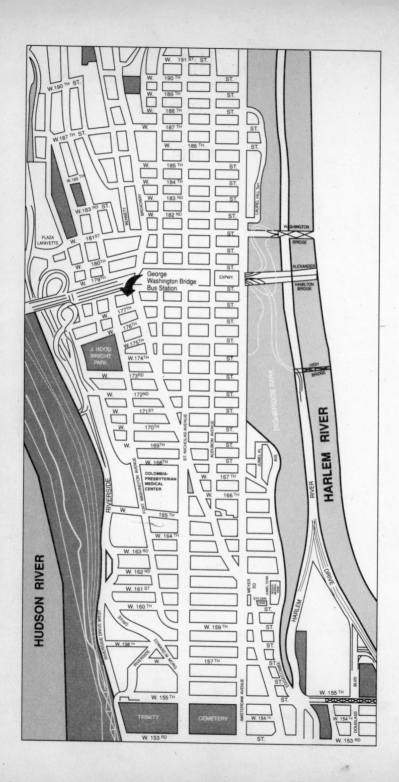

The
Cocaine
Kids

The Inside Story
of a Teenage
Drug Ring

Terry Williams

Addison-Wesley Publishing Company, Inc.
Reading, Massachusetts   Menlo Park, California
New York   Don Mills, Ontario   Wokingham,
England   Amsterdam   Bonn   Sydney
Singapore   Tokyo   Madrid   San Juan

Library of Congress Cataloging-in-Publication Data
Williams, Terry Moses, date.
    The cocaine kids : the inside story of a teenage drug ring / by Terry Williams.
        p.   cm.
    Includes index.
    ISBN 0-201-09360-X
    1. Narcotics dealers—New York (N.Y.)—Case studies. 2. Narcotics and youth—New York (N.Y.)—Case studies. 3. Cocaine industry—New York (N.Y.)—Case studies. I. Title. II. Title: Inside story of a teenage drug ring.
HV5833.N45W55 1989
364.1′77′097471—dc19   89-6505
                                                                                       CIP

Jacket photograph by Edmundo Morales
Jacket design by Copenhaver Cumpston
Text design by Joyce C. Weston
Set in 10 1/2-point Primer by Interactive Composition Corporation, Pleasant Hill, CA

ABCDEFGHIJ-DO-89
First printing, June 1989

This book is dedicated to
William Kornblum

Contents

Preface ix

Acknowledgments xi

Introduction 1

Doing Ethnography · Changes in the Cocaine Culture · The Distribution Network

1. The Setup 13

The Crew · The Neighborhood · El Cubano · A Day in the Office

2. The Cocaine Trade 31

The Consignment System · The Base Craze · Hector's Fall, Max's Rise · Risks of the Trade · Cocaine Buyers · Cocaine Houses · Cocaine-amatic · Scramblin in the Street · Jake's Troubles

3. The Kids   63
      Kitty and Splib · Splib · The Birthday
      Party · Irene and Splib · Kitty and
      Carlos · School Days · Max and
      Suzanne · Masterrap and the Slang

4. The Scene   93
      Jump-Offs · The After-Hours Clubs · The
      Hierarchy · Staying on Top · The Crack
      House and Base House Trade · Kitty and
      Dial-a-Gram

5. The Kids Move On   117
      The Shooting of Chillie · Splib · Hector ·
      Max · Jake · Charlie · Masterrap · Kitty's
      New Life

Afterword
                                                          131

Glossary
                                                          135

Index
                                                          139

Preface

This is a story about teenagers who move in a very fast lane, each one trying to be "the king, making crazy money for as long as I can, any way I can."

It focuses on the lives of eight young cocaine dealers in New York City. From 1982 to 1986, I spent some two hours a day, three days a week, hanging out with these kids in cocaine bars, after-hours spots, discos, restaurants, crack houses, on street corners, in their homes and at family gatherings and parties.

These studies took me to the Bronx, Harlem and Washington Heights—areas of high unemployment and diminishing resources, especially for young people. But while quality entry-level jobs were disappearing, illegal opportunities were emerging with considerable force because of the growth of a powerful and profitable multi-national drug industry.

In describing the role teenagers play in the illegal drug market, this book touches on a number of problems now much discussed, including drug misuse, the mechanics of distribution, and the sexual behavior of drug-users and drug-sellers. But these are not the true essentials of this account. At the heart of this book are the kids' own stories of their lives; of their struggles with family problems, high incomes, girlfriends and boyfriends, with running a business and, finally, with making decisions about their own futures.

My intention is to throw light on a major and complex social problem, but without blaming the victims and without placing teenagers in stereotypical roles. Every teen aspires to make good. In the cocaine hustle, that means to "get behind the scale"—to deal in significant quantities; it is like landing a top sales job in a major corporation, or being named a partner, after a long apprenticeship, in a brokerage firm with a seat on the Stock Exchange. The kids who get that far have some control over prices and selling techniques, direct the work of subordinates, and, above all else, make large amounts of money.

The teenagers in this story are sophisticated cocaine distributors, wholesalers and retail sellers. Their work has been essential to the growth of a major industry; they have helped establish an organizational structure that sustains a regular market and outwits law enforcement authorities. These teenagers have also found a way to make money in a society that offers them few constructive alternatives.

In many ways, these kids and others like them simply want respect: they are willing to risk their lives to attain those prized adult rewards of power, prestige and wealth. Theirs is a difficult and dangerous way of life, one closed to almost all outsiders. Because they must continually make tough decisions in trying circumstances, these young people grow to adulthood with little time to be young.

Acknowledgments

It was my good fortune to come to know the young people whose lives are portrayed on these pages. If they had not trusted me, guided me through the maze of their world, shared their insights, given me their total support, this book could not have been written. I would like to thank them here in the names they chose for themselves: Max, Charlie, Kitty, Splib, Jake, Masterrap, Hector, Suzanne, and Chillie. For obvious reasons the identities of the Kids and some of the settings have been disguised, and every person quoted or described in the book bears a name not his or her own. A special thanks to Jane Ciampa for her knowledge, support and general assistance during the many years of research. She has been a guiding force for me and my work.

I am especially grateful for the assistance of Sergio Vasquez, Aris Vidal, Victor Montgomery, Iris R., Chicky, Ralph, Evette Tally, Shawn Smith, Caramelia, Julio, Akemi Kochiyama-Ladson, Tselane Williams, Miriam, Alkamal Duncan, Edmundo Morales and Ali Smith. I thank Angela Gaviria for her knowledge and assistance in translation.

I owe a particular debt to my friend and colleague Bill Kornblum for the intellectual vigor he imparts, his unselfish and irrepressible encouragement.

I thank Margie Thybulle for her indefatigable typing of the manuscript and more importantly her ideas, thoughts, criticism and unique judgement. Thanks also to Carol Barko, my editor and friend, whose energy, keen eye, wonderful personality and graceful skills helped make this book possible.

Being a staff member at the Conservation of Human Resources at Columbia University from 1986 to 1988, where I worked on a Ford Foundation-supported study of four metropolitan areas, facilitated the task of revising this manuscript for publication. Special thanks to Dr. Eli Ginzberg, director of the Conservation, whose encouragement and enthusiasm helped make the writing of this book so much easier. I greatly appreciate the assistance of Conservation staff: Charles Fredericks, Thierry Noyelle, Thomas Stanback, Anna Dutka, Thomas Bailey, Shoshana Vasheetz, and Ellen Levine.

Field work for this book was made possible by a grant from the National Institute on Drug Abuse (NIDA) when I was a post-doctoral fellow at the New York State Division of Substance Abuse Services. I am grateful for the generous encouragement and collegial support provided by Drs. Charles Winick, Bruce Johnson, and Greg Falkin, my mentors in this extremely valuable program to assist minority researchers. A good friend and present member of that program, Alisse Waterston, assisted me early on when the manuscript was only a series of snapshots, giving her time freely to read and discuss the possibilities of the work.

This book would not have been possible without the support, patience and encouragement of my family: my sons Kahlil and Neruda, janice williams, Sandra Smith, Donald Smith helped me in innumerable ways, encouraging me to complete the book; thanks to my brother Michael for his strength and support by example. My friends Lucille Perez, Philipe Bourgois, Sandy Close, Vernon Boggs, Claire Sterk provided moral support during the writing of this book; I hold them all in the highest collegial regard.

I am pleased to express my appreciation to Dr. Eric Wanner, who provided valuable comments on the final draft of the manuscript despite a very busy schedule.

I want to extend my most sincere appreciation and heartfelt thanks to Jane Isay, my editor at Addison-Wesley, whose editorial assistance, keen scholarship, and enthusiasm made the whole enterprise worthwhile.

Finally, the value of having an editor-as-detective is no small matter and Peter Solomon is owed this distinction for his careful reading and teasing out of the many confused parts of the Kids' dealing world. I especially acknowledge an indebtedness to him for his editorial skill and sharp-eyed word mastery.

I hope this book meets the expectations of those who helped me along the way, but in the end I hold the responsibility for its shortcomings.

"My crimey here thinks the way to go is more drugs. But I know better. I think making money is okay, but not making it just by dealing. You gotta go legit, at least for a minute. You gotta go state of fresh, all the way live, if you wanna do anything worthwhile out here. Everybody thinks they can make crazy dollars, but they confused. It ain't like that. I've seen co-caine bust many a head—they get fucked up and be clocking out after they find out they cannot find the key to understanding the mystery of skied. I say skied. But you know what? But-but-but you know what? They don't have a clue. Word."

Masterrap

Introduction

The apartment is crowded with teenagers, all wearing half-laced sneakers and necklace ropes of gold. Doorbells ring every few minutes, white powder dusts the table tops; jagged-edge matchbook covers and dollar bills seem to flow from hand to hand. The talk is frenetic, filled with masterful plans and false promises. Everybody has a girl. Everybody has cocaine. Everybody has a gun.

This is a book about kids, cocaine and crack. But it is also a book about work and money, love and deceit, hope and ambition. It describes—as much as possible in their own words—the world of teenage members of a cocaine ring: the way they do business, their neighborhood, their families, their highs and lows.

When I first came on the teenage cocaine scene, I was apprehensive, even fearful. I knew these young people were volatile and unpredictable, prone to violence, and not inclined to trust adults—or, for that matter, anyone outside their circumscribed world. Yet I wanted to find out about the kids who sold drugs. How did they get into the cocaine business, and how do they stay in it? How transient is their involvement—can they get out of the business? And where do they go if they do? What are the rewards for those who succeed?

The only way to find the answers to these questions was to follow the kids over time, and that is what I did. For more than four years, I asked questions and recorded the answers without trying to find support for any particular thesis. In the process, I found that the truth was embedded in a complex, miniature society with institutions, laws, morality, language, codes of behavior all its own. I also found young people whose only shield against fear and uncertainty was a sense of their own immortality.

Doing Ethnography

This attempt to provide a rounded and dynamic portrait of the kids, their work and their world, relies heavily on a form of research known as ethnography.

An ethnographer tries to describe everyday behavior and rituals and, in the process, to reveal hidden structures of power. As this technique requires the researcher to build a close relationship with those being studied, it is necessarily slow: days, sometimes weeks, may pass before the ethnographer can even begin to conduct an interview. These interviews are often "open"; that is, the investigator has key questions in mind, but is willing to let an informant's responses lead into unanticipated areas as these can provide new understandings of the processes under study.

Ethnography also involves careful observation of individuals in their own social setting, and systematic recording of their action and speech. This can include simple quantitative measures, such as noting the sex, age, and ethnicity of participants, or observing a particular routine. For example, in observing cocaine transactions in a bar or apartment, I might count the number of buyers and sellers; record the techniques used to make (or conceal) sale or use of the drug, the prices paid and the purity claimed; even tally the number of times patrons visit the bathroom, often a favorite place to snort cocaine discreetly.

But ethnographers also record far more subtle information, such as use of language, gestures, facial expressions, style of clothing; and must watch with care to capture exceptional episodes that can be particularly illuminating.

Clearly, detailed and descriptive field notes are essential in this approach, especially as the observer makes every attempt to accurately record the speech of those observed. In my work, this was often very difficult, even after the kids freely welcomed me into their world. For one thing, any tape recording made in the midst of the turbulent business of cocaine would have been unintelligible—there are phones ringing, people coming and going, and often a more or less constant background of family arguments, babies crying, loud music, and other disturbances. In addition, taking handwritten notes during a conversation warps discussion and inhibits the flow of words.

Besides, although the crew fully accepted me and my work, others—their customers, the hangers-on, cocaine groupies—were not necessarily informed about my role: things were fine as long as I was seen as another waiting buyer or a friend; producing a pad or tape recorder would certainly have stirred suspicions. Another compelling reason to avoid behaving like an observer was the potential for violence. With guns openly visible and police raids a real possibility, I felt I had to keep my hands free, my eyes sharp, and my mind clear.

For these reasons, I developed a method of jotting down key words or phrases immediately after each visit, and reconstructing conversations or a scene from those notes the next day. (There were a few exceptions: I did take sketchy notes of some private, one-to-one conversations with individuals I had known for years.) It was not unusual to spend a day or more writing up an hour or two of field observations.

Despite such obstacles, more than 1,200 hours of field work produced six thick notebooks, including drawings and diagrams covering every nuance of the kids' operation—methods of production and packaging, forms of dealing and the flow of cash—and a great deal of material about the structure of dealing networks and the rituals of cocaine use.

Although I became close to the kids and people in their world, even grew to think of some of them as friends, some important distances could not be breached: for one thing, they were teenagers; for another, while cocaine use is a routine part of their daily lives, I never consumed any drug stronger than alcohol, and did not participate in any way in the preparation, transport or sale of drugs.

Changes in the Cocaine Culture

In the years I spent with the kids, there were major changes in the cocaine trade: in the ways the drug was prepared and distributed, in the preferred styles of use, and in prices. This book details the effects of these changes on the kids and others who work at the middle level, providing the essential link between big-time importers and small-time users; here I want to offer a brief overview of the patterns in the larger cocaine culture as background for the stories that follow.

Cocaine is obtained from the leaves of the coca plant. It is an "alkaloid," one of a large group of substances found in plants. They are usually bitter, and often have drug-like effects: caffeine and nicotine are alkaloids; some, such as quinine, are regularly used for medicinal purposes, often in synthetic form (for example, the dentist's Novocain is a type of procaine, a synthesized form of cocaine).

As a pharmaceutical preparation, cocaine is usually prepared as a white crystalline powder, cocaine hydrochloride. Hydrochloride is in effect a salt which is added to the alkaloid; the compound dissolves readily in fluid to make an injectable solution, and cocaine hydrochloride was once widely used as a local anesthetic.

In non-medicinal use, the drug in powder form can be

inhaled like snuff (this is called "sniffing" or "snorting"), mixed with tobacco or marijuana (smoking), or injected ("shooting"). All methods produce, more or less quickly, a sensation that is described as intense, ecstatic—and very short-lived, lasting only a matter of minutes.

It is generally agreed that users do not develop a "tolerance" to cocaine; that is, do not need to ingest ever-larger amounts to obtain the sought-after effect. However, the desire to repeat the experience can mean that a single user will do everything possible to obtain and use as much of the drug as he or she can.

In the cocaine culture, methods of use have changed over time. At the turn of the century, extract of coca was widely available in tonics, wines, and teas, until the drug was outlawed in 1922. Most early accounts (including those by Sigmund Freud and about Sherlock Holmes) involved injection, and the 1950 edition of the *Merck Manual*, a popular ready reference guide for physicians, informed its readers that almost all illicit use of cocaine was through intravenous injection—often repeated every few minutes. "Pure cocaine addiction now is almost nonexistent," the guide went on, because, in large amounts "the effects are so unpleasant."

In the 1960s and most of the 1970s, snorting was very much the preferred style. However, by the late 1980s a relatively new form had taken hold: some 80 percent of those who use the drug are now smoking it in the form of "base" or "crack"—terms for the basic alkaloidal level of cocaine. This can be made simply by boiling cocaine in water; the residue is placed in cold water where it forms an odd-shaped, off-white, hard mass. Pieces of this mass—the base—are then chipped off; the substance makes a crack-ing noise when it is smoked.

When base (or "freebase") first appeared in the late

1960s, users rolled it into marijuana joints and cigarettes; some used the residue (called "'due" or "con-con") in the form of an oil to saturate joints. These practices preceded the more elaborate forms of "freebasing"—smoking the base, sometimes mixed with other narcotics, in pipes or in a device made from laboratory pipettes—which arose in the early 1970s among cocaine connoisseurs who believed, mistakenly, that reducing cocaine to an alkaloidal level (removing the hydrochloride) would make it "healthier," like natural food enthusiasts seeking to free themselves from any adulterated product.

In part, the growing popularity of base reflected deliberate moves at high levels in the cocaine hierarchy. By 1978, importers had stockpiled so much of the drug that they feared prices would drop unless demand increased strongly. So they began to include a small quantity of base in kilo shipments as a "gift" and urged dealers to entice buyers to experiment with the "new" product. This did not find immediate acceptance; it was quite expensive, for one thing, and basing was at first restricted to a small group of big spenders. But around 1980 or 1981, use of base did begin to catch on in New York City.

Dealers were at first reluctant to offer the new product because they actually made more money on the powder— the process of removing the impurities from the drug left them with less weight to sell. With powder, impurity was the key to profit: cocaine brought high returns because it could be adulterated ("cut") several times. Many dealers also objected to spending the additional time required for processing, and as customers demanded purer and purer cocaine, dealers turned to suppliers and pressed them to provide unadulterated material in kilo quantities for the first time.

Then in 1983, outside forces intervened again. A glut

on the market in the producing countries (Peru, Bolivia, Chile) forced foreign suppliers to cut their prices. This had an immediate, beneficial effect for dealers, who continued to charge their retail customers the old price and realized extraordinary profits for a time. Word did get out eventually, but even at lower prices, sales could not keep up with production and total dollar receipts were down.

At this point, a new product was introduced which offered the chance to expand the market in ways never before possible: crack, packaged in small quantities and selling for $5 and sometimes even less—a fraction of the usual minimum sale for powder—allowed dealers to attract an entirely new class of consumers. Once it took hold this change was very swift and very sweeping. By 1984 or 1985, only a few customers were asking for powder.

One other extraordinarily volatile element is the value of the commodity itself. Over the years of this study, prices for cocaine changed so often and so drastically that it is impossible to write of the dollar value of a transaction in any consistent way. For example, a kilo (2.2 pounds) of cocaine worth $50,000 in 1980 brought only $20,000 in 1988, but this general downward trend included many sharp ups and downs along the way. A complicating factor is that price changes of the raw material at the wholesale level often have little or no noticeable effect on street prices.

In addition, there is strong incentive to inflate dollar amounts involved on every side—not only by law enforcement officials who overvalue seizures, but by dealers at every level who want to impress others in and out of the trade—which makes it impossible to provide any sensible dollar estimate of the trade as a whole. For all these reasons, prices are quoted only rarely in the course of this book, and then only to illustrate the situation at a particular point in time.

The Distribution Network

At the retail level, the distribution and sale of cocaine in New York City involves mostly African-American and Latino boys and girls under eighteen. In general, they come from families whose income is below the poverty line, and from neighborhoods where there is little chance to rise above that line. It is difficult to say how many young people are engaged in this trade, but certainly there are many thousands in the metropolitan area.

Many teenagers are drawn to work in the cocaine trade simply because they want jobs, full time or even as casual labor—the drug business is a "safety net" of sorts, a place where it is always possible to make a few dollars. Teens are also pulled by the flash and dazzle, and by the chance to make big money, and pushed by the desire to "be somebody."

Those who recruit teenagers are following a tradition that dates back almost twenty years, and was the direct effect of the harsh "Rockefeller laws" mandating a prison term for anyone over eighteen in possession of an illegal drug. This led heroin dealers to use kids as runners, and cocaine importers have followed this pattern: young people not only avoid the law but are, for the most part, quite trustworthy; they are also relatively easy to frighten and control.

TODAY, teenagers who work at the retail level are expected to sell cocaine and crack for cash in a way that generates repeat business—in other words, to act like sales people in any business. In addition, they are expected to limit their own consumption of the drug, keep accurate records, and avoid arrest. Most of the kids are users as well as dealers, and so cocaine rather than cash has often been the medium of exchange; the introduction of crack has changed this, as most dealers disdain the use of crack, so cash payment is now more common.

Cocaine is a highly valued commodity, especially among the middle class, but because distribution and sale must be clandestine, reaching users on a regular basis presents problems. Thus there are important roles in the network which do not directly involve selling at all. For instance, there are "runners"—messengers who take cocaine to buyers or let buyers know of a particular dealer; the runner earns a "p.c.," a part commission or percentage of the sale from the dealer. Where the drugs are sold from a fixed location there are "lookouts" and guards, and often "catchers" standing by in case a police raid or other emergency means drug stocks must be moved swiftly. At this level, there is considerable flexibility, and an individual may shift quickly and easily between several roles: today's lookout may be tomorrow's runner; a door guard may eventually move to a selling position.

At the wholesale level, "transporters" move large amounts across state lines to prearranged locations, where a "babysitter" may keep watch over them. Import arrangements may involve "swimmers" who retrieve packages from the ocean and "mules" who transport (sometimes unknowingly) quantities into the country.

For clarity, I use distinct words for the various roles involved, even though there is no consistent terminology on the street. On these pages, "dealing" refers generally to selling any amount of cocaine; "distributors" provide wholesale amounts of the drug, usually in kilogram lots, to "suppliers" who in turn provide smaller amounts to "sellers," who make actual retail transactions, exchanging cocaine for money. Processing or adulteration may take place at any level.

Suppliers are critical actors within the distribution system. In one sense, they function somewhat like jobbers: importers offer them merchandise in wholesale lots; they are responsible for supplying retailers with that merchandise. In another sense, they are like middle managers: they must hire, and supervise the work of, several full-time employees, and a number of on-call supporting personnel.

Sellers are not only involved in retail trade; they also transmit a certain amount of lore regarding the rituals of cocaine use, and provide free samples. Despite this, they are often disliked or distrusted because they determine the purity and set the price of the product. They are also feared because they are usually armed or accompanied by armed guards.

The emergence of base and crack has set in motion major changes in this marketplace: lower prices, discounting as a selling strategy, broader availability, crack houses; it has also increased police pressure on consumers, led to new forms of official corruption, and brought a much higher level of display—and use—of weapons in cocaine-related business.

Money and drugs are the obvious immediate rewards for kids in the cocaine trade. But there is another strong motivating force, and that is the desire to show family and friends that they can succeed at something. Moving up a career ladder and making money is especially important where there are few visible opportunities.

THE name "Cocaine Kids" is my invention; they have no such name for themselves. "Tags [names] is for kids who write on the walls of the subway," says Max. "This is serious business here. Word." But they do use words in special and important ways. In the world of the kids, slang can be seen as a form of social criticism, with an emphasis on shocking or confusing people from the outside while amusing friends. It is also a way of confronting their own anxieties by poking fun at the idea of growing up.

I have tried to record their speech accurately because their new and modified words and phrases speak to their present lives in ways that our standard vocabulary does not—at best, they provide a rich, even musical, background to their everyday world. Each term is defined when it first appears, and there is a brief glossary at the end of the book. The Kids themselves are generally comfortable speaking either Spanish or English. In my presence, they spoke English most of the time, though occasionally—in anger or in intimacy—they preferred Spanish.

1

The Setup

Max was fourteen and already considered a "comer" in the cocaine business when I met him. We were introduced by a Dominican friend who knew of my interest in New York's underground cocaine culture and the teenagers who survive outside the regular economy. I assumed we would talk and then go our separate ways: he trusted my friend but he was shy; there was certainly no reason for him to talk with me about anything, and I was not about to press the issue. But there was something special about Max, and he became my friend and guide for nearly five years. I think we got along because I was an outsider and he had a story to tell, and he chose me to tell it to.

The day we met, Max introduced me to two young African-American kids, brothers, Chris and John. One was a college student, the other had left high school during 10th grade. Max had been trying to set up an organized "crew" of teenagers to keep his business going, but so far his hopes had been dashed by ineptitude or betrayal. His troubles with these two were to prove typical of the problems he encountered when he first tried to establish a crew.

Chris and John lived on 143rd Street, a key location for Max's cocaine trading because it is close to a major subway stop at 145th Street, has a strong mixture of African-

Americans and Latinos, and is easily accessible to the heavy traffic that flows along Riverside Drive to and from New Jersey. It was a place to "cop" (buy), a "copping zone." As they sat in the apartment Max explained his price structure, the kind of cocaine they could expect, and his delivery schedule.

The brothers were personable and intelligent, and we talked about girls and sports as well as discussing drugs in general. Most interesting was to hear Chris on the psychopharmacology of cocaine—he had ambitions of becoming a pharmacist, and discussed the varieties of cocaine and what it would look and taste like combined with certain chemicals. He was particularly interested in the effects of new types of adulterants.

It was two or three months before Max saw that the brothers were not working out. They had been giving the cocaine away to girls, partying, showing off with friends and otherwise doing things that were not good business. Max had to begin the selection process all over again. This time he wanted to be more careful, but he was still not exactly sure how to do that.

He eliminated one part of the problem by cutting back on the amount of cocaine the brothers received at one time. He would also telephone to ask for "call money" about mid-week, requesting payment before it was actually due. This is a common practice with new crew members, but it became a key feature of Max's modus operandi for all he supplied.

To avoid hearing the kids say they had been robbed or busted or had lost the cocaine—or any other excuses—Max learned to pick up money due to him with little or no notice especially when he knew the dealers had been partying too much. Having too much money around, he felt, made it easier for the kids to "fuck it up" so he would take whatever cash was on hand.

He also learned it was better to go and get the money rather than wait for others to bring it to him. With new dealers, the amount on hand when he appeared did not matter so much, as long as they had some money to show they had been on the job. Not to have any money when Max called might signify that the dealer had loaned out too much cocaine or, worse, was not hustling to sell the cocaine on hand. Young recruits who did not hustle not only reflected poorly on the supplier's judgement, they could damage his cash flow—and leave him unable to meet his commitment to upper level importers.

Max's visits also served to let crew members understand that they could not be in business without him. As he moved up in the business, he modified this procedure and relied on threats—of violence and of withholding cocaine—to coerce young dealers into complying with the rules. Once new recruits gained his confidence, he was willing to wait a reasonable time for them to call him. This was a sign that he thought they were mature and had made a major step toward establishing trust, a factor of no small significance since much of cocaine dealing is predicated on trust.

Standing not quite five feet ten inches tall, with dark intense eyes, Max was considered taciturn, a reserved young man who confided in few people and revealed himself to no one. When I told him I wanted to write about him and about teenagers in general, he made no effort to conceal the everyday workings of the business, and actually made it possible for me to witness events I would never have seen without his approval. He introduced me to his friends and family (his brother Hector and his aunt) and I was able to follow their lives for nearly five years.

We lost contact from time to time, but he often got in touch so I could witness events he thought would help my research. For example, when he felt a cocaine shipment was unique—"Peruvian flake," for instance—he would call to ask if I wanted to see it. There were other special occasions, such as the time he introduced me to his fiancée, Suzanne, or when he started to manufacture and process base for the crack market and wanted me to see how it was done.

I believe Max trusted me for two reasons. First, I insisted on telling him the truth; second, I never revealed anything he said to me in confidence. My discretion was important not simply because of the danger he faced if his business became known to the wrong people, but because he is a private person and had been betrayed many times: by his own brother, who took cocaine and never returned the money; by girlfriends who used him because he had money, cocaine and jewelry; and by others he thought he knew well. Trusting a complete stranger was risky, and Max at times tested me, telling me some things just to see if I would repeat them to those in the crew. I never did. He would often say, "Don't tell this to anybody" or ask, "Did you say anything to anybody about what I said to you?" Telling the truth was as important for me as it was for him. I had a great deal to lose if he thought I was deceitful.

Over several years, Max assembled the crew called here the Cocaine Kids. He introduced me to them—Chillie, Masterrap, Charlie, Hector, Jake and Kitty—at "*la oficina*," the office. This is an apartment, rented by Chillie (though not in his own name), where the Kids cut and mix cocaine, pick and pack crack; it is also the base from which they sell unpackaged cocaine to individual buyers.

In the summer of 1984, with Max's approval, I had a chance to observe the operation at *la oficina* which was run by Chillie, the crew boss and Max's first lieutenant. There was constant rivalry between all members of the crew, but I had to get along with each one, and that would not have been possible without Max to pave the way. They knew I was recording their lives and they accepted me and let me know their secrets, professional and otherwise. I did have something to give them—functions to perform, roles to play; small things, but they meant a great deal to the kids: I was kind of a big brother, able to help with homework and even babysitting, but most of all a willing and sympathetic listener.

The Crew

Everybody sits around as Max prepares the crack. Max is a master at mixing. He uses his own recipe and is familiar with the effects of the drug in various combinations.

Jake snorts from a large bag of cocaine resting on the glass table, telling Chillie about a woman he met at Jump-Offs, an after-hours club. "*La jeva no era muy grande, pero tenía lo de atrás* [The girl was not too big, but she had a big behind]." He kisses his fingers. "*Hombre, 'mano.*"

Max's recipe calls for an "eighth" of cocaine (1/8 kilo, or 125 grams), 60 grams of bicarbonate of soda (ordinary baking soda) and 40 grams of "comeback," an adulterant that has allowed Max to double his profits from crack: this chemical can be cooked with base and, when the base is dried, it smells, tastes, and looks very much like cocaine; all that is used "comes back." At $200 an ounce in 1984, it cost far less than the real thing.

He fills the Pyrex pot with tap water, and sets it on the stove to boil. After 20 minutes, he places the material in cold water to coagulate into crack, and members of the crew come forward to cut the hardened chunks with razor blades and pack the chips into red-topped capsules.

Hector

Hector is skinny and freckle-faced. He is only three years older than Max, but he moves with the gait of an old man. His hands are rough, his bloodshot eyes dart from object to object with a twitchy nervousness. Though he was once a major dealer, and in many ways Max's mentor, he now looks to his little brother for support during hard times, and today is one of those times. He will only admit he made a mistake when he is high and "the cocaine is talking."

"I know I fucked up and made some vicious mistakes when I was behind the scale, but it's not gonna stay that way. I'll be back. Everybody is entitled to one mistake. I used to handle the weight [pounds, half pounds, kilograms] and I still can. It ain't no problem."

But Hector is not a kid anymore, and he knows it. As he talks, nobody looks up. When he asks, *"Mira, loco, si tu quieres, yo te puedo ayudar con eso* [Hey man, if you want me to, I can help you with that]," Max does not answer. Hector's eyes focus on Jake's hands counting crack chips into the capsules.

Jake

Jake is rotund, and always wearing faded jeans, dirty unlaced sneakers, a soiled T-shirt and a bummy-looking leather jacket. He is the odd man in. He looks older than his seventeen years and is sometimes shy. Jake has been sniffing until it's time to pick and pack the white chips. He says cocaine gives him courage to face the unpredictable street.

"Did you hear about Max's uncle?" he asks so quietly one can barely hear him. Max interrupts, tells him to get some foil to wrap some powder. Jake straightens up, gets the foil, and places it on the table in front of Max. Again, he mentions Max's uncle, but nobody pays much attention. Jake seems as if he is about to explode—it turns out he badly wants to know if Max's uncle was killed because he owed a dealer for crack. Max does not answer.

A tireless worker, honest and loyal, Jake would never hurt Max. "I never lived in New York until my mother brought me here," he says, in a tone that is almost apologetic. He met Max by accident in the street. "He told me to go to this spot with him. He said he needed some back [backup or help] and he didn't have anybody. I went with him and made the move OK. After that we go back to see Chillie—Chillie used to have this spot in the Bronx then.

"After this happened we come to Max's house and he asked me to go to work for him. I knew him because my sister knew his wife. Then we find out we're kin because of my aunt. I like working for him because it's easy money. I just watch myself and then I don't worry about too much."

Chillie

Chillie is the boss at *la oficina*, which means that he supervises the work of Masterrap and Charlie. He and Max are the same age, and have worked together three years. He has dark, wavy hair and a sneaky smile that rarely surfaces; when it does it gives him a handsome but sinister look. He lets everybody know that he, not Max, should be controlling the cocaine business because he takes in more money than anyone in the crew except Max. "I made over a million dollars selling this stuff. If the connect [connection; the importer] knew what I was doing, he would want to see me. Max knows I do the best business out here. I don't want except a little money and a little respect." But Max won't introduce him to the Colombian supplier.

Masterrap

Masterrap is slick, articulate and cool. He is quick to inform you that he is a ladies' man. "Rap is my name, females my game." When Chillie is busy or out of the office, he takes over; he is the second man behind the scale. While he does not appear ambitious, he is the first in the crew to see the bigger picture and is well aware that time is against everyone in the cocaine trade. He does not overindulge like the others do.

Masterrap has his heart set on a musical career and has written many "rap" songs that he hopes one day to record. "Coke is just a way for me to make some money and do some of the things I would otherwise not have the chance of doing in the real world. Coke ain't real. All this stuff and the things we do ain't real. If I told you half the things that go on in this place you wouldn't believe me. I wanna tell you my life story one day, and after you put it down I wanna see it and maybe then I'll believe this is really happening and not a dream."

Charlie

Charlie is the only African-American on the crew. He is a bodyguard at the office; he and Chillie were high school friends, and now they are partners. He has taken three martial arts courses and learned how to shoot a gun after his uncle—a New York City corrections officer—took him to a rifle range.

Charlie is eighteen and looking to be the next man behind the scale.

Kitty

The only woman in the crew, Kitty is five foot six, mulatto in complexion with high cheekbones and an engaging smile. She is depressed today; says she is tired of the cocaine-dealing hassle. "I really just want to go back to some school a few times a week. It would be better if I went

Monday, Tuesday and Wednesday because then I would have time for my kid"—her son Armando, two years old.

(The kids often say they want to get out of the business when they are depressed—or when business is poor, after a bust, when there are family problems or lovers' quarrels. But when things are going well, they are eager to be out on the street trying to make a dollar.)

Splib

Splib, Kitty's husband, is the only person present who deals as an independent, though he sometimes functions as a member of Max's crew. At nineteen, he is the oldest here. He is wise, handsome, and above all else a survivor. He also takes great pleasure in his ability to con and manipulate people.

For an example, Splib proudly tells about his birthday, when he took two ounces of cocaine and two women to a hotel in New Jersey. They stayed three days and three nights. Afterward, broke and depressed, he called his supplier and told him he had a buyer who wanted four and a half ounces—then worth $5,000—right away, explaining that the two ounces were "on the street" (being distributed), but were not yet paid for. His supplier agreed to meet him with the cocaine, and Splib was able to sell enough to pay for the two birthday ounces as well as the new consignment.

"I never worry about money," he snaps. "I can always make money." Excited by his own story, he takes a folded dollar bill from his shirt pocket, and opens it to reveal what he announces to be the "purest cocaine in the world." Bending the edge of a matchbook cover into a vee, he takes two quick snorts. Refolding the bill with one hand, he is now ready to go into the street.

Splib, like most of the crew, is Dominican, but his ability to speak both Spanish and African-American slang

with facility, and to mingle in both worlds is a valuable asset
to the operation. He is aware of this, and high-handed about
it: "The Indians [Colombians] have so much coke they can't
off [sell] it without finding new markets. Blacks have proved
they can organize and sell the shit, but the Indians don't
know how to deal with Black cats. They don't understand
their world or the way they do business. I do. And they know
it."

THE crack is packed in vials, the powder allotted. Max tells
Jake when to return for more. The money is to be dropped
off at another location. Chillie and Kitty get their consignments; Charlie and Masterrap take one last snort before
they depart. Everybody is ready to deal.

The Neighborhood

Washington Heights stretches north on the west side
of New York City from about 154th Street to 190th. In 1957,
when Max's parents moved into the Heights, they were one
of the first Dominican families to do so. Broadway, which
runs north and south, was a dividing line: few Latinos and
no African-Americans were allowed to rent apartments east
of Broadway; few whites lived to the west.

It was truly a color line, says Mel, an African-American
who works at a garage on 153rd Street. "You had those light,
damn near white Cubanos who lived here and there during
that time. Now we call Broadway 'Dominican Avenue.' This
area was all Jews, some Irish, and some Italians. They had
a couple Spanish clubs like the Pan American Club on
157th Street and Cabarrojeno on 145th. Some Spanish
would come to the clubs but it didn't mean they come into
the neighborhood to live, because clubs is one thing and
living in an apartment around here is something else."

For a month or so in the fall of 1776, General Washington's headquarters were here, at about 160th Street, until the British overran Fort Washington. "I remember when they built that crematorium down at the bottom of Trinity Cemetery, and nobody said a thing about it. But you know they had to be digging up the graves of colored soldiers who fought in the Revolutionary War because there were so many of them buried in unmarked graves," says Mel.

"Nobody seems to have any history no more, not the parents, not the kids. All the kids around here are Dominican." He looks at the parking lot, which contains a high percentage of BMWs. "All these fancy cars they're driving they couldn't afford without the drug money. I don't know for sure it's drug money, but I'm not stupid either."

Certainly it is true today that, although many different ethnic groups call Washington Heights home, Dominicans shape the character of this vibrant community. On a summer afternoon, the streets are teeming with Latino noises, smells and talk. Men—restaurant workers, street hustlers, store clerks, maintenance workers—gather to play dominoes or cards or shoot dice in the shade of a barbershop awning or a faded sign above *la grocería*. There is Astroni's, always serving breakfast, lunch and dinner: young dealers and customers go there to make phone calls; police officers drive up in the small hours for coffee take-out.

Saturday morning around 9:00 is a quiet time for the neighborhood drug trade, but vendors of other street merchandise are busy spreading their wares on the sidewalk and most of the regular businesses—Julio's Head Shop, Charlie's Metro Bar, the Greek-owned coffee shop, and many eateries—are always open. In the Monarch Bar, you can tell the dealers by the beepers clipped to their belts and by the way they handle money: they don't simply take out one bill to pay, but display the entire wad, counting off the bills in rapid strokes.

Along 156th Street, the house numbers signal the

entranceways to many of the cocaine and crack houses in the area. Dealers like Jake stand cocksure in the doorways with their hands across their chests, luring customers— they are only a short walk from the subway stop—into the hallways. "Gypsy cab" drivers who have stopped to drink beer and snort a little *perico* talk near the barbershop; some are playing a favorite local card game, *veinte-y-uno*, twenty-one. With money stacked high and every player eyeing the cash, they slam cards down with a cry loud enough to stop any passerby.

The games are strictly for men. Women get special attention in other ways: whenever a young woman with a substantial front or a protruding behind passes, the men look up to whistle, suck their lips, and blurt out, "*hola mami* [hey, baby!]," (or "*hola negrita* [hey, dark sweet thing]" or "*hola gorda* [hey, big momma]"). Women less well endowed are apt to hear less admiring, but no less Dominican, responses: "*Eso no es nada que sabor* [that's nothing but a 'taste']" is an often-heard comment. Most women ignore these remarks, but some of the teenage girls on the block parade by regularly, slowing gracefully until the men take notice, then strutting away.

Micro-enterprises fill the street: men and women of all ages hawking shirts, hats, pocket calculators, watches, radios, handbags, jewelry, food, all without protest from storekeepers who seem willing to grant their *compadres* an equal opportunity to make money. A worn graffito across the base of a nearby building reads: GOD BLESS AMERICA AND THE YANKEE DOLLAR, and that seems to be the creed underlying the economic bustle of legal and illegal businesses.

Illegal transactions are a leading activity here, and hustling is the name of the game. One sign of this is the large number of secondary operations catering to the drug trade: candy stores that sell drug paraphernalia and head shops that sell little else, such as Perran's, across the street from one of New York's most active copping zones, its display window filled with an assortment of lactose, dextrose, mannitol (all used to adulterate cocaine) and a host of water pipes for smoking crack or marijuana. Perran's is open 24 hours, and, like many businesses in the area, is thriving.

Less visible are bootleggers who sell a local version of corn whiskey popular among cocaine users, the cocaine bars frequented by local dealers and users day and night, and the after-hours spots, a community institution, that also survive on the cash spent by cocaine habitues.

Cocaine also occupies a central place in the lifeways of this community in another, invisible sense: Salsa, Latin jazz or the Spanish language have never received wide acceptance in the United States, but cocaine surely has. As coca and cocaine are, after all, Latin in origin, there is some nationalistic pride mixed with the Dominican and Latin entrepreneurial drive here.

Yet in a way drug dealing is no more the focus of this community than the recent influx of Korean-owned businesses. This is a lively multi-ethnic community with family-run *grocerías* and liquor stores, check-cashing agencies, banks, restaurants, discos, hardware stores and so on. Boricua College, the newest institutional and cultural addition to the neighborhood, is the most recent tenant in the landmark McKim and White building at 155th Street, which also houses the American Indian Museum and the Hispanic Society as well as the Numismatic Society, The Geographic Society, and the Society of Arts and Letters.

Across the street, the Church of the Intercession, the oldest church in Washington Heights, has programs for youth, for teenage pregnancy, AIDS prevention, and a strong interest in making this a better place to live. Nearby, Esperanza (Hope) Center, a community-based organization, helps new arrivals adjust to life in New York City, assists in landlord-tenant disputes and family problems, helps find jobs for youth and provides basic education for adults.

But here, too, despite the protests of many residents, teenagers and adults alike have taken to crack in unprecedented numbers—as in other cities, large and small. Washington Heights, which the police call a "hot spot," is a battleground in the war on drugs. And as in all wars, it is the young who are the first casualties.

El Cubano

It's 6:30 Friday evening. El Cubano, a noisy little dive near 158th, sports new curtains and a bright Michelob sign. The establishment is no more than 20 by 40 feet, with a small bar and six close-together tables. Two video games, a poker machine, a jukebox and a cigarette machine give the place an amusement center atmosphere which is amplified by the mirrors on three walls: a long one behind the bar, another against the wall leading to the lavatories, still another behind the jukebox.

The owner and a few friends often play cards in a tiny back room that also serves as an office. Nearby are two bathrooms, tiny unkempt cubicles. As women seldom frequent the bar, the men are as likely to piss and snort in one as in the other. The few women are looked over carefully. Young girls will come in to use the phone or the bathroom, but only a certain type (only whores, according to Jake) come to drink alone.

The owner, a Puerto Rican with family ties in Cuba, winks at cocaine snorting by regular customers—almost all of them hustlers who buy drinks because the cocaine makes them thirsty. The owner, Jake says, "tells us not to snort at the bar, but we do it anyway." In general, however, the patrons are not too blatant. "He say the bathroom be chill for that—he don't want us to get caught in his place, so we be chill." Not so long ago, underage youth were not allowed to come into El Cubano. But as the ranks of hustlers filled with kids under eighteen, the owner found their fast dollars hard to resist and today teenagers make up 75 percent of the bar's business.

Jake has several "war stories" about El Cubano. A favorite is the one about two cocaine-crazed women who raped a young barmaid when she went to the bathroom. Jake claims he was in the bar when it happened, but he did not actually see the encounter. Max says the three girls knew each other and were all Lesbians anyway. Another story is about a shootout between two dealers fighting over a woman who snorted one dealer's cocaine and not the other's. The spurned dealer, enraged, fired several holes into the jukebox.

There's a story about that jukebox, too. It is a classic Wurlitzer with 1950s rock, blues and vintage Afro-Cuban jazz records and only a few contemporary tunes. Jake says the owner fell in love with an African-American singer who loved these songs so much that he had the machine locked so they could not be removed. After he and the singer parted company, he did nothing to change the box. On the day of the shooting, when the owner returned to the bar, he walked over shattered glass, ignoring the broken chairs and tables and other debris, straight to the jukebox to make sure the records were intact.

A Day in the Office

La oficina is located right in the middle of this bustling, mixed Washington Heights neighborhood. It has a large steel door specifically designed to prevent the police from knocking it down before the kids could dispose of the drugs. Chillie has hired the fourteen-year-old son of the building superintendent as a "catcher"—he is on call to retrieve any cocaine thrown out the window during a bust. The stock or "stash" of cocaine is kept in a bag stitched with beads worn by adherents of *Santería* (a set of religious practices involving Roman Catholic and African elements), and the boy's parents will not touch it because they fear it contains evil spirits. Chillie pays the boy in cash and cocaine.

The office is a small, one-bedroom apartment. A newcomer who enters the living room will see only a sofa, two chairs, a stereo, a TV, a tiny stool and some plants. Business hours are usually 1:00 PM to 5:00 AM, six days a week. The three office workers are Charlie, the armed door guard; Masterrap, who acts as host and receives requests from buyers; and Chillie, the man behind the scale, who sets prices, arranges to barter goods and services, gives credit and makes day-to-day decisions regarding sales.

At *la oficina*, unlike many "coke houses," the scales, packaging material and the drug itself are not immediately visible. Chillie prefers to deal from the thickly-carpeted bedroom, with a full-sized platform bed, a desk, and a large walk-in closet filled with candles, coconuts, beads in water-filled jars, coins in large bowls, cigars, and a silver plate holding food and money. The desk and telephone, the center of operations, sit near one window, facing the door. In the middle of the desktop are aluminum foil packets evenly cut to wrap the cocaine, and a triple-beam scale. No cocaine is visible until the buyer has shown his money.

Chillie arranges for me to sit where I can observe clearly. He says I can come into the bedroom with some buyers, but not all, as some are "funny about that." "If anybody ask who you are, tell them it's none of their mother-fuckin business. Just play past that shit." And for the next several hours, I sit in the living room with customers, talking with Masterrap or watching TV, until Chillie calls them. Few customers talk with Charlie—as the guard, he refuses to engage in any conversation while he is on duty.

Once in a while, the buyers are friends, and Charlie or Masterrap will negotiate with Chillie on their behalf. Chillie pays a bonus for every buyer they bring in, ranging from a few dollars on gram or half-gram sales to $100 for a one-ounce sale.

The phone rings constantly. Chillie answers from his room, and in a few minutes the doorbell rings. The first buyer today is a young Spanish woman who speaks halting English. She calls for Chillie to come out and meet her; they embrace and kiss. He asks how much she wants to buy and they go into his room. The slide on the scale makes tiny noises as it is moved along the ribbed bar. Ten minutes later she departs, after (he tells me) giving him $60 for a gram.

The doorbell rings again. Masterrap gets up and hollers "Back" to Charlie, who walks over and stands behind him. Masterrap peers into the peep hole; "it's cool," he announces, and Charlie relaxes and goes to sit in the kitchen. Two teenage Dominicans come in, wearing sneakers, leather jackets, and gold chains, and brand new blue jeans. Masterrap goes into a rap about music with them, then goes in to get them a taste from Chillie. They snort a bit, "take a freeze" (place a pinch of cocaine on the tongue), chat another minute or two, then go in to see Chillie.

After they leave, Masterrap and I talk about a movie we've both just seen. Charlie is in the kitchen, eating and yelling commands at a dog who is tied down. The dog, an akita ("They used to guard the emperors of Japan"), has a long curled tail, a strong face, and, though attentive, does not appear vicious or at all concerned with the goings-on. "They be mean dogs if you train them right," Charlie insists, lifting a piece of bread so the dog will jump for it. "Hey, they are better than them pit bulls out there," he asserts as if looking for an argument. When the dog moves about, Charlie shouts, "*calmate, calmate*," and it sits, head held high. Some time later, Chillie shot the dog because it ate two ounces of cocaine mistakenly left on the table.

In all, I see fifteen buyers come and go that afternoon. Each one tastes the cocaine before purchasing; none stays more than twenty minutes. The first buyer is the lone woman. Masterrap explains, "You know we have plenty of females coming in here not just to cop but to hang out. Chillie don't go for that all the time. We gotta limit it. If things are slow, we let them stay longer or we might call a freak [a girl without inhibitions] to come over."

Buyers would say how much they wanted to purchase, and, after learning the price, would ask for discounts. Most sales were in the $60 to $80 range (1985). After the last customer, Chillie came out of the bathroom crowing, "I am The Deal-Maker."

2

The Cocaine

Trade

Max is relaxed, settling down with a glass of beer and a joint. "I'd rather smoke reefer than sniff because I'm tired of sniffing all the time. I gotta cut down anyway." His wife, Suzanne, comes out of the bedroom to tell him he has a phone call. Max prefers to be called on the tiny beeper attached to his belt. If it is business or urgent, he will call back from a public phone. The beeper, a constant companion, is also a constant source of annoyance; if it beeps when he is busy, he frowns—but if the number registers, he will, more times than not, answer right away.

As in any business, Max worked his way up by impressing a superior. He wanted to prove in one gesture that he was a good risk, thrifty, had good taste and was not ostentatious. He accomplished all this with a friend's automobile. "My friend had a Benz. A big one. I told him I could get crazy drugs if I had that to front off. I wanted the big Benz because a lot of my friends have the baby Benz but the big one is what the big time is all about.

"This was when my connect was having problems with his people. He put me on with this other connect but he was afraid to give me anything big, you know anything over what my regular connect had been giving me. So one day I borrowed the Benz and went out to see him. Well, when he saw me drive up in the Benz, word, all of a sudden his attitude changed. And when my connect came back I was making crazy dollars for this new guy and my connect was pissed off—he started offering me crazy prices to think about taking his package again."

Max's street reputation helped him impress the first Colombian importer to supply him with larger amounts of cocaine; then over time he proved to be a good credit risk. Finding reliable teenagers is vital to drug suppliers at all levels: they assess a user/dealer's behavior over an extended period to find those who can be "loaned" a certain amount of cocaine and will return the correct dollar amount on time. "Nobody trusted me with any material at first," Max says. "I had to convince people I could do it. I didn't have my hand out for no charity. I worked hard to get established."

Some suppliers require that a new dealer be recommended by other street hustlers, but despite all precautions, suppliers frequently encounter people who are always a day late and a dollar short. The problem is compounded because, unless the dealer makes a major mistake, it is very difficult for suppliers to know if he or she is showing signs of overindulging or otherwise becoming unreliable.

Chillie tells of what he believes was a test of his reliability. "I had just started with Max and he wanted to know why I was late on paying him for a package. I told him why—and it was the truth—but I could tell he didn't believe me.

"About a week or so later he called me and said he wanted to hang out with me and that we could call over a couple of females to have a good time. Now, he had just left here and all of sudden he calls, he wants to have a good time. What's up? I thought, Something's up. We went to his apartment and he had these two freaks there. I mean freaks.

We were sniffing all night long. It was chill. At about five a.m. they wanted to base and shit and I knew Max didn't base either. So when they brought out the pipe I just waited till they finished; then we fucked all over the place. After that Max musta thought I was cool because it's been years now that we're together. He just wanted to know if I based and if I'd lied to him about it. I know he was just checking me out."

Before freebasing became popular, a dealer might have assessed recruits on the basis of other standards, looking for signs of self-indulgent behavior such as flashy attire, or an obviously expensive cocaine habit.

Those who prove to be bad credit risks ultimately have trouble finding suppliers who will give them cocaine to sell, but they are not completely excluded from the distribution system. Many are taken on in a variety of tangential roles and work as steerers, touts, guards, runners, and "cop men"—dealers whom suppliers will only sell to on a cash basis. Of course, even reliable individuals can lapse into overindulgence now and then, or may be robbed, and these circumstances must be tolerated to some degree. But, as Max underlines, "If one of my people say they got busted I wanna read it in the papers. If it ain't in the papers somebody's gonna get hurt."

A persuasive enforcement mechanism is the threat or actual imposition of violence upon those who breach trust and abuse the privileges accorded dealers. Most suppliers must eventually resort to threats of violence against some sellers to get appropriate returns, and such threats must sometimes be carried out. Sooner or later, this hits close to home: "My uncle was killed by a dude who used to cop from me," Max says without emotion. "He [the killer] is a big time dealer now with keys [kilograms] and shit, and he shot my uncle seven times."

The Consignment System

Except for Hector, who is not trusted, crew members who sell to retail customers work on a consignment basis. Various terms are employed to describe this: suppliers may say they give cocaine "on credit" or as a "loan" to distributors; at all levels, it is called "fronting."

Each week, Max is fronted three to five kilos (in 1985, this would have a street value of $180,000 to $350,000) to distribute to the kids. The quantity he receives varies according to the quantity he has previously sold, how much he has on hand, and how much he is committed to deliver both to the crew and directly to other customers.

In his turn, Max supplies each crew member with the amount and kind of cocaine, crack or powder, needed. Each kid is then responsible for selling his or her share in a designated time; the kids return either money or cocaine to Max. He then pays the supplier or returns the cocaine, although returning the merchandise is frowned upon.

The quality, variety and amount of cocaine each crew member receives, then, is determined by Max and by his suppliers. The Kids might ask for more but whether or not they get more depends on how much they sold from their last consignment. Personal factors are also involved. Once Suzanne overheard Chillie criticizing her housekeeping to Max and suggesting that she was lazy; angered, she complained to Max and he was honor-bound to show his displeasure, so he cut back on Chillie's consignment package and raised the price.

"The price is going up, because the connect wants thirty [$30,000] for each [kilo] package, and I got to make at least ten on each package for myself," Max explained. "The Colombians control the whole thing. If they say thirty, it's thirty. If they say forty, it's forty." And this is true enough: the Colombians, as importers and controllers of the supply, do set the price. But Max is also rationalizing: once the drugs are in his hands, he has considerable latitude, and can use that to get back at Chillie.

He also has considerable latitude when it comes to the quality of the product once it leaves the importers' hands. One day he gives me a demonstration, emerging from the back room with "something I just got"—a plastic bag containing four chunks of cocaine about four inches in diameter. Each piece is individually wrapped in paper towels, secured by two thick rubber bands; each weighs 250 grams.

"This is fresh stuff," he says, taking a few of the flaky particles and rubbing them between his forefinger and thumb. "I don't do nothing to this. I leave it like it is in case a customer wants to buy a whole kilo.

"If it's a new customer, I might take thirty grams of flake and cut a little of it," he explains. In other words, he might remove 30 grams of the uncut drug for his own profit or pleasure, and replace that with adulterant. "But you can't take more than sixty grams of flake if a guy is paying forty thousand dollars." (Despite this lesson, Max often acts as if adulterating the product is beneath him: "I leave the cutting to my people downtown. They handle the gram business.")

Kitty says Max has changed his mind about how much "shake" should be in a kilo. Shake is the term for a mixture of cocaine powder with adulterant: most suppliers will allow up to 120 grams of shake to a kilo, or 12 percent; kilo-level buyers are usually unhappy if they find more. There is also disagreement on how much of a kilo should be "flake off the rock" as opposed to the more valued crystalline form. The flake-like shavings occur naturally as the drug is transported but many buyers complain if they see too much.

Rarely is any one member of the group allowed to distribute more than 250 grams—a quarter kilo—at one time. A crew leader like Chillie is given half that, an "eighth," each week. He decides how many grams to loan out to his workers, Masterrap, Charlie, or Kitty. In the same fashion, crack dealers are consigned so many capsules: Jake, for example, is allowed 100 to 150 capsules, selling at $5 each, at one time.

Virtually all cocaine suppliers expect retail dealers to return at a specified time with cash amounting to about 60 to 75 percent of potential retail sales of their consignment. Say Chillie receives from Max an "eighth," valued at $3,000. Chillie charges exactly the same price per gram as he paid, but he has cut the cocaine by half: the drug he sells his customers is one part adulterant to one part cocaine. As he has transformed 125 grams into 250 grams, his gross sales should be $6,000 for a profit of $3,000. In practice, he may use a portion for himself (or for friends), and street prices can rise or fall very quickly, so he may take in more or less than $6,000. Max gets a percentage of the total retail sales: typically, Chillie would be expected to return $4,000 in a week, a profit for Max of $1,000.

Chillie says he always gets his cocaine on consignment. "I ain't gonna pay for it first. You don't do business like that. If the connect wants to off it, he has to come to me. He needs me. So, I take it and off it—if I'm a little short, I replace it the next time." He produces a baggie filled with yellowish powder, "all flake—stuff off the rock" and points to the white powdery lactose that he "whacks" (mixes) it with to get shake.

I ask if Max gets upset when he, Chillie, is late or short on cash. "Max never gets anything short from me. I don't work that way. If I'm short I don't say. I just give him what I have and that's that."

I press him, "Suppose you get a pound, and you owe him cash for a pound—if you take him less than that aren't you short?"

Without missing a beat, he slips away from the question. "No," he says, "it don't work like that. First of all, I don't get no pound all at once from nobody. I don't need that much coke around here. I take an eighth and I whack up two ounces [56 grams] of it, and sell the whacked stuff by the gram, half-grams, like that. I leave a few rocks in it. I keep some of it for my head [personal use], and I turn on a few people who buy nice from me. I sell the rock with some shake to people who buy quarters [1/4 ounce or seven grams], and pure rock only to my very best customers."

Chillie regularly complains about his supplier: "Max gives me a quarter for my head and sometimes I whack that up but I mostly keep it for myself. He should give me more—I make him a hundred thousand dollars a year and I don't see why he can't give me more. But he knows I'm gonna take some anyhow.

"He promised me he was going to get the big thing [a kilo] and I told Kitty to tell him I'd be waiting. I told him I didn't have no front, and he said it was OK.

"Max knows there is a lot of money out there and you can't have your package short [underweight] like as far off as the last one he gave me. Masterrap and me played with that package forever, we worked it to death [cut it as far as they could]. It was short. Max says he didn't put it on the scale, he eyeballed it. I said 'bullshit'; you know and I know nobody in this business eyeballs no big package like that, especially Max. People weigh a half gram out here, so you know he didn't eyeball it."

Chillie's refusal to admit that he owes money to Max illustrates a common problem for street distributors: they can easily owe more money than they can actually make. The process of wheeling and dealing, learning how to make money—and how to control one's own consumption—is an ongoing one. Some street dealers make no money at all one year and are bringing in "crazy dollars" the next.

During his years with Max, Chillie moved up consider-
ably, and was given larger and larger amounts of cocaine to
distribute. But the path was not always smooth, as when
Max—responding to Suzanne urging him to fire Chillie for
insulting her—chose to punish Chillie by raising his price.

"Max was charging five hundred dollars more than I
could get in the street," Chillie said. "He's crazy. So I went
to my cousin." Max knew very well what prices were, but he
was confident that other suppliers could not provide cocaine
of comparable quality on a steady basis. Max's judgement
proved correct: Chillie was back receiving his usual consignments after only a few months (and Masterrap and
Charlie, who had been lying low, returned to the scene,
relying again upon Chillie for their supply).

In mid-1985, Jake and Kitty were receiving cocaine on
consignment from Max, but Splib—who was then loosely
associated with the crew—was being cut off. Splib owed
Max money, and Max was refusing him any additional
cocaine. "Splib," said Max, "thinks he's gonna re-up on my
stuff when he gets the money he says is owed him in the
street. If he knew what he was doing he wouldn't have that
much money in the street." It took Splib some time to
straighten things out with Max.

Cocaine dealers like Max and Chillie assign a relatively
fixed retail price per gram after they adulterate the drug. As
in any business, this price is designed to produce a profit;
that is, if all the cocaine is sold at the set price, the dealer
would have a given amount of cash money in excess of what
he paid. But dealers often make the mistake of considering
the money they actually receive to be their own real returns,
ignoring the fact that most of this money is owed to the
supplier or must be spent, as in any business, to achieve
maximum sales.

The question of value is further confused by the fact that dealers are often expected to provide each buyer with a free pinch of cocaine to show confidence in the drug's purity or quality. Dealers also tend to take from their regular supply when they entertain at bars, after-hours clubs, and parties. True, suppliers often provide dealers with extra cocaine for these purposes (dealers say it is rarely enough; suppliers say it is more than sufficient), but once this "calling card coke" is depleted the dealer and friends begin to snort up profits; that is, to use their inventory not for business but for pleasure and often all the cash money they take in must go to the supplier.

Because cocaine can be used in many ways—injected, snorted or smoked—dealers adulterate, process and reprocess it into any number of saleable forms, looking for innovative ways to maximize their profits. This very variety makes it difficult to calculate weights and purities with any precision. And while dealers like those on Max's crew proudly claim to have "the purest cocaine in the world," anonymous street buyers rarely know what they are getting. (This is less true of base and crack, which can be assayed by any of several simple procedures.)

The Base Craze

By the winter of 1983, pure rock was the craze much as crack was to be four or five years later. Many users had grown tired of sniffing cocaine, and sought out the best and most efficient high, which—they were convinced—meant "rock" they could cook into freebase for themselves. Rock is the term for the supposedly "pure" crystalline form of cocaine, the highest quality available.

One response to freebase buyers' increasing demand for purer and purer cocaine was a proliferation of dealers and con men ("beat artists") purporting to sell the real thing. To meet that demand, dealers in turn began to call for pure rock with no flake from suppliers. Dealers like Max also learned to "manufacture" rocks. For example, filling a plastic baggie with cocaine and blowing air into it, then squeezing the bag tightly, will create small rock-like nuggets of cocaine, giving buyers the impression that they are getting pure product though in reality the drug has been cut more than it normally would be.

This was called "recompressed" cocaine. Kitty explained. "Recompressed is like when you take coke and put some cut into it and make rocks so the person who buys it thinks that it's pure, but actually it's been cut." Another method, "reconstituted" she said, "is the same thing except you gotta do it with a machine. But you can tell if something is reconstituted because it's usually too hard. If coke is old, not very fresh, it's like Rice Krispies—'snap, crackle, and pop'—well, if it's reconstituted the same thing might happen."

Masterrap and Charlie both noted the sudden change among their friends: "I had a female friend who came to me one day and asked if I knew anybody who sold coke," Masterrap remembers. "She didn't know I worked at the office. I told her, 'Yeah.' She said, 'We wanna get some rock, pure rock,' in this squeaky little voice. I told her all that glitter ain't gold and she didn't know what I meant. Everybody all at once wanted rock. I talked to Chillie about it and he just said, 'It's gonna cost 'em,' with this big smile on his face."

"I told two of my old roomies about the coke house, but I didn't say I worked here or nothing," says Charlie. "Steven, I been knowing him a long time—he's really a smart kid, he's into physics and stuff—and he asked me did I know somebody who could get rock. He said he was taking out these girls and wanted some rock coke—he told me it was important to get rocks to impress the females."

Hector, who has had his share of cocaine in all manner and variety, does not think much of all this. Standing erect, his arms across his chest, he makes a gesture with an index finger pointing straight out and the thumb up.

"Yo. When people say they want rock, most of 'em don't know what they mean. People who deal know. When I dealt the weight I'd see all kinds of packages. The kids out here don't know a flake from a fish—if you asked them what fishscale is, they wouldn't know. [Fishscale is high-grade cocaine powder with few rock-like chunks.] Some of the best cocaine comes in flakes, not rock, like Bolivian flake—right, Max?—and the Peruvian stuff we got long time ago, that is the best cocaine in the world and it's not rock, not like rocks. You know, by the time it gets here with all the banging and shipping it's all flake.

"Rock coke is just a crazy thing people got into themselves that it's the best there is, but it ain't. I know there is good blow out here that is rocks, don't mistake me, but the stuff they doing to get the rock into the street is not the real thing.

"When I was selling weight my consumers automatic got so much flake and so much rock. Automatic. They didn't say, 'Give me all rocks.' They knew this was the best stuff, and they knew how the rock falls off and makes flake."

For Max, the rock craze translates into dollars. "Most of this thing about rock is really just because basers want more for their money, and regular sniffers do, too. It's hurting me because people don't want to pay two hundred dollars for a gram of pure rock, but they wanna base up the best stuff. I'm getting lots of customers who now only want that base; they don't want the rock no more, they want the stuff already cooked. Now, I lose when this happens because all that don't come back [the amount of cocaine lost in processing] I have to take as a loss.

"Before, when I sold them the cocaine and they cooked it, they took the loss. Something's gotta give. I got to talk to my connect because the shit is getting funny."

Max recognized the trend early on, and directed the crew to supply the best cocaine in the neighborhood, hoping to get a larger share of the street market. But despite Max's directives, Kitty, Splib and Jake altered the contents of their packages to deceive buyers, other dealers—and each other—as often as they could.

Splib, for example, continued to profit by cutting already adulterated cocaine. He is a most creative mixer. "I only put lactose on the *perico*," he says with a grin. "But the way I do it is special. This half ounce [about 14 grams] has been whacked once and I can sell it as it is. But check this out. Look carefully at these pebbles." He took an extraordinarily fine strainer, obviously finer than the one used by his supplier, and sifted the powder leaving three and a half grams of small pebbles.

"Now, I take these and put a one on it [add an equal amount of "cut"] and I make seven grams. I leave a little bit of these pebbles out. Now I got about seventeen grams. I take eleven grams of flake and put four grams of cut into that and I got twenty-one grams. I take two for myself and I off the other nineteen, pay Max back his five hundred dollars and take the other fourteen hundred dollars for my slide [pocket].

"I'm gonna put lactose in here. You see, when you heat the stuff [lactose] it takes some of the sugar content out, and then it makes the coke look fluffy and fresh. You notice I didn't take all the pebbles away. That's because when I finish cutting it, I sprinkle some of them over the package so everybody will see a little shine when they open the foil. I always put this kind of product in foil, because in after-hours spots where I sell it, they don't see so well and a few shining pebbles are cool.

"Also this coke don't stick to the foil like it does with the heavy lactose, because it ain't that much sugar in it. I don't put but a few twennies [$20 packets] in foil anyway 'cause if you sweat too much it cakes up. I prefer to carry a baggie and scoop it out cause that way they think you're in the big time."

By the end of the summer of 1984, only the most naive buyer would tolerate cocaine with any adulterant. Base buyers probably constituted 70 percent of the Kids' clients—once production problems were resolved—and represented a very profitable repeat business.

Hector's Fall, Max's Rise

Hector called me aside one day. "I would like to say something to you about what happened to me," he says out of the side of his mouth. He is edgy, animated, always on the verge of some tense move: biting his nails, pulling his nose, pacing, going to the bathroom, crossing and uncrossing his legs.

"You know, certain people put you in a position, but they don't have no business being there. I mean like the cops—they don't know how to go about dealing with people. When I got busted for taking some coke in my car, I admit I was incorrect in that move. But the cops never told me about my rights. After that happened, I just wanted to strike back in some way. All my life I have seen people like that, who hold you in a position to tell you what to do, or keep you down. And they want you to give up trying."

Max, for his part, viewed his brother's problem as mostly caused by Hector's own ignorance and obstinacy. Max said Hector figured nobody could hurt him if he did what he was supposed to do, and knew he was capable of doing, and so Hector wasn't worried about getting out of any situation. Max has his own version of Hector's story:

"Well, my brother was moving big stuff—half a kilo weekly—by the time he was sixteen. He was showing everybody he was Mister Big Time. And he was. He was taking girls upstate for weekends with ounces on him, he had a BMW all souped up, he had gold—too much gold. Every night was a party.

"Then one day he didn't have nothing. He wasn't taking care of business. People be telling him nobody be staying out front for too long. If you fronting, you gonna take a fall. Anyhow, some dudes robbed his apartment while he was gone on one of them weekends with females. I bet they set him up—I told him they did. He lost a couple of ounces in that one. He lost a few thousand dollars, too, and they took his weapons. This caused problems with the connect, you know, because he was tired of hearing stories, he wanted to get paid.

"Then things went okay for a while, until the police saw him leaving a spot one night and followed him a few blocks, said he ran a red light. He had on lots of jewelry, he had two females with him and shit, and they found two ounces on him, about two thousand cash dollar bills and no driver's license. He never had a driver's license—he was too young to get one. He went to jail for a year.

"After that, he wasn't no more good. He had been death on basing before he went to the joint—he would always say to me that people who based was weak and simple-minded. But when he got out he was basing all the time. In a few months everybody could see he had a big problem. I mean a big problem. He was still making some money for his connect, but he owed him twenty thousand dollars. The cops took the car. He wasn't getting no respect from his people because he wasn't reliable enough and plus everybody knew he was on the pipe. It was killing him.

"My aunt and cousins, my uncle and me, we all knew we had to do something. After we got together, we decided to send him back to Dominica [the Dominican Republic]. When the connect came by to get his money, I gave it to him after putting everything together and selling some of the gold and shit. The big gold plate [neck piece] that was mine, I let Hector take it home to D.R. with him. The main thing was to have the connect's money. He asked what happened to my brother and I told him. He felt bad and told me to keep five thousand dollars to help pay for all the things I had to pay for. Then he asked me if I still had the customers [list] my brother had and I told him I did. He said if I wanted to keep working where my brother left off—and I did—I would have to get behind the scale."

Unlike Max, most dealers start out in lowly positions, then move up through hard work, skill, intelligence and a little luck. A kid who can routinely handle money, control personal use of cocaine, deal with buyers, and control a weapon, may make it out of the street and into the elite world of the super dealer. Getting behind the scale is the first real step—indeed, the major move up in the cocaine business. As Masterrap put it, "It is just a way of saying you wanna take some control and make more money. Everybody wants to get behind the scale because they can make that money, buy clothes, a baby Benz, and impress the females."

Max runs part of his operation in a loosely organized way, hiring workers as needed to sell cocaine and crack in a variety of locations. He can draw from a sizeable labor pool of teenagers who have dropped out of high school and are unemployed. Like workers in the above-ground economy, the kids have a chance to be promoted and make more money; many can look to an eventual move up behind the scale.

Risks of the Trade

Cocaine selling requires mobility, careful planning, and swift action. Max has used motorcycles, limousine services, messenger services and private cars to deliver and distribute cocaine. "I use the motorbike because I can get around fast and I can throw the stuff away if the cops follow me. If I got to take a big package, I never use my car, because the rule is, 'if you use it you lose it.' I don't use messengers any more because they might split with a package if they find out what's in it."

It also requires consistent caution; there is a sting of danger to every transaction, and sellers become expert at concealing the drug until they are sure they are dealing with a genuine user, not a police officer. Kitty says she always carries the coke in her bra when she is on the street; in a limo, she stashes it under the seat: "the windows are smoked so nobody can see in there and plus cops don't bother to stop limos anyway."

When she responds to a call from a prostitute whose customer wants the drug she has to worry about the "trick" being a cop, so she asks the women not to beep her until "the guy has already propositioned himself. Cops can't do that—that's entrapment. But a lot of these guys don't want sex, so it gets kind of tricky."

Those who work from a fixed location like an apartment are in a particularly dangerous position, as they are literally trapped, and must develop a strategy to avoid arrests and robberies. Chillie knows, "I'm the one that's gonna take the fall if this place is busted. The steel door is only so much protection. I gotta be able to get this stuff out of here in a hurry; I use Peppi [the super's son] to catch some of the stuff, but he ain't around all the time. I know the cops go to the back stairs sometimes when they bust people so that may not work for very long.

"But the apartment is not in my name, and I'm moving in a way that the whole thing is pre-packed: I'm gonna put all the cocaine into foil before the customer comes in here, I'm gonna make up all the twenties, the fifties [$20 and $50 packets], the whole thing so there won't be no hassles and no sitting around sniffing and shit. That way I won't have to have this scale no more."

(This effort did not survive in the market. Customers complained that the aluminum foil packets caused the drug to melt, and thought the amount was short; they began to demand their money back. Max, hearing of this, found waterproof packets, called snowseals, to replace the foil.

(But many customers were still upset that they could no longer "taste" the cocaine, long a part of the sales ritual— and especially upset because Chillie still allowed some of his friends to sniff before they bought. Eventually, the complaints mushroomed to the point where he had to go back to using the scale and the old method of doing business.)

All the Kids snort cocaine regularly. This is accepted, but the use of crack is generally frowned upon: those who snort are thought to have more control and discipline than those who smoke crack or freebase. Most dealers see crack smokers as obsessive consumers who cannot take care of business; crack users, they say, tend to become agitated, quickly lose control and concentration, and take one dose after another at the expense of everything else. Snorters, however, can use the drug and still take care of business.

Yet the Kids who snort do so on almost any pretext: Chillie will sniff and ask Masterrap and Charlie to join him whenever he makes a sale of more than $100. One day they decided to celebrate because the New York Mets, their favorite baseball team, won a game. They called Max, began snorting and playing music, and called girls to join them.

On less joyous occasions, cocaine serves a therapeutic purpose, as an antidote to stress, disappointment, and the problems of everyday life.

The Kids need no excuses—they can use as much cocaine as they wish—but they apparently feel the need to explain their behavior. When Masterrap missed an appointment with his girlfriend he came back to the apartment and said he wanted a "hit" because the girl was "messing him around." As Masterrap prided himself on being in total control of all his girlfriends, this was an unlikely admission.

Cocaine Buyers

Max boasts of having cocaine available 365 days a year. This day is Friday, May 30, 1986, about 5:30 pm. We are in an apartment he has rented in the South Bronx; for furniture there is one chair, two old sofas, and a bed, nothing more, and only the telephone is new. "Call me back at about six o'clock," he tells a caller. Friday is not a day to be without cocaine—the weekend beckons—but at 6:00 pm Max is still without cocaine to sell.

"I have coke three hundred sixty-five days a year," Max repeats, "and I don't like to be empty when people call. Right now, I'm waiting for my man to come through; it's taking him a little longer tonight because we made a deal on a new price and he has to talk to his people. It's better to let the customer think you are home and busy than to think you are out of coke."

Because it is bad business to be without cocaine, a dealer will often let his phone ring rather than answer and tell buyers there is nothing to sell. "No businessman likes to say 'no' to a possible sale," Max argues. "If they think you are out they will go to somebody else. But if they know you can get coke and you have a rep for good coke, they might wait."

Most cocaine dealers attract customers on the basis of their reputation, street word, street runners, and other sources, including newspaper or television stories that pinpoint a copping zone. Chillie relies on Masterrap, Charlie, Max, and many friends—a word-of-mouth network—to steer buyers to his office. "My friends," says Charlie, "are always asking me where to buy cocaine and crack. I always tell them about the office. And if they want me to bring it to them, I do that." Jake says, "I meet people on the street. They say they like coke before they say they like crack. Max tells me to tell him about the people who want coke and I do. Sometimes I tell Chillie too."

"Once the office was open," Chillie says, "I never had any trouble getting customers." Masterrap recalls the early days: "When we first started. I used to stand downstairs and tell customers where to come if they wanted something. I was down there in the snow, in the rain. Now, we got phone numbers and beepers and shit. You've come a long way, baby," he says to Chillie who cracks one of his rare smiles.

"I always meet people who want coke because they have heard about how good my package is," says Kitty. Dressed in a blue silk dress, instead of her usual casual clothes, she looks more like a woman than a young girl. She sometimes works as an in-house dealer, that is, she holds an exclusive right to sell in a particular place, with the owner taking a percentage; other spots are basically open to any dealer.

"You know, in this business the better the coke the better your chances of making money. But in some of the spots I worked, it didn't matter how good your package was so much as just having something all the time. Never be without, that's what it's all about. If you had a good package you could sell it, but if you had a bad package and you were the only one with a package you could sell it. The bad package is the customer's last resort, but if you got nothing else, what are they gonna do? Most customers want good coke but not all of them can afford the best stuff. But customers know you and want to see you; they want something personal."

Max says, "My customers come from other dealers, when they run out of material. Splib and Kitty can tell you about Anthony. Anthony was a big-time dealer and he had this customer buying heavy shit and he runs out and sends him to me; the guy never went back to Anthony. I also get customers from my people downtown, who tell me if somebody wants to get into any weight.

"But most of my customers now are people I've known for a long time and who have made it to the weight level. You know what I mean? I don't wanna meet nobody now who comes to me and says, 'I wanna buy a key.' I have to know who it is. Or I say to whoever brings the customer, 'get the money. If they trust you with the money I can get the material.' Kitty got me a good customer who wanted to buy a pound, and I did just that. Splib has gotten me some good customers too. Even my brother got me some customers.

"The thing you got to remember is my customers are not that many. They get a pound or a key every week. They call me if it's late, but most of the time I call them to tell them where to meet me and how much to bring if they owe me from a previous package."

Splib's customers, for the most part, come from the street but (always manipulating) he claims "they might be lawyers, doctors, teachers, everybody. At first, I had only hustlers, numbers runners, other dealers, bartenders, barmaids, you know, people like that.

"One time a dealer friend of mine ran out of coke and called me to meet him at this hotel downtown. I brought about four grams for this chef friend of his to try—he said he knew everybody: show business people, actors and shit. I thought he was lying but my man said he was telling the truth. He paid me for the coke and told me he wanted more, but basically he was not my customer, so I just sold him the stuff until my man got back on his feet again and the romance was over.

"Most of my customers now are street people, night people. Working people buy once or twice a week; on the weekend. They start buying around Thursday night and this lasts till Sunday. After Sunday, only the hustlers buy."

Cocaine Houses

Before Max established his present crew, he tried dealing from apartments rented with friends and acquaintances who came in with cash as partners. Many of these partners did well for a time, but could not keep it up—they stole money, "lost" cocaine, made too many mistakes, like those first recruited for the crew. Some lasted longer than others, but fellow Dominicans have proved to be his most consistent partners.

This seems to be true in the cocaine culture as a whole. Dominicans are the foot soldiers for the Colombians, and while there are many young African-American, Puerto Rican and Colombian teenagers transacting cocaine business, Dominicans are by far the majority among dealers who operate cocaine and crack houses in New York City. Dominicans were in charge of 50 of the 53 coke and crack houses I visited.

Because of the intense competition in street trafficking, and the lack of available housing in Latino communities in Manhattan, Dominicans began setting up businesses in predominantly African-American communities like Harlem around 1983. They were able to gain a foothold because they had a large pool of available cash from cocaine sales, could tap into a large African-American cocaine-consuming market—and because there were so many abandoned and semi-abandoned buildings in these neigh-borhoods.

Another crucial factor which allowed Dominicans to move in was quality. African-American dealers are notori-ous for selling over-adulterated cocaine, while Latinos, be-cause they are closer to higher-level suppliers, can sell a relatively uncut product. The introduction of rock cocaine and more sophisticated methods of adulteration have some-what increased competition, but there has been remarkably little violence between African-American and Dominican dealers.

Between 1984 and 1988, cocaine distribution increas-ingly came to resemble a fast food chain. Cocaine and crack houses proliferated, offering low prices, fast service, and good quality. Most houses operate more than one office in a single building, which allows for high-volume purchasing by buyers who need not be referred by other buyers. During this time, prepackaged units, selling for $20, $40, $60, $100 became the standard. With adequate backup—lookouts, guards, touts and stash catchers—Max says these opera-tions are low risk.

There has been considerable community opposition. Marginal areas, especially those on the verge of gen-trification, are most active in attempts to eliminate or at least control drug activity, either by calling for police intervention or by taking organized actions, such as demonstrating at or near crack and cocaine houses—one group in New York marks the entrances of crack houses with red paint. So far, public protests have had no demonstrable effect on the distributors' operations.

Dealers prefer rundown, abandoned buildings in poorer neighborhoods because their tenants and landlords tend to be apathetic. Indeed, one important ingredient in the ability of drug-dealers to expand is the availability of abandoned and semi-abandoned properties in poor sections of the city such as Harlem, the South Bronx, the Lower East Side, Bedford-Stuyvesant, Jamaica (Queens) and Washington Heights. Landlords have all but left many buildings for dead in these areas, and dealers either rent several apartments on a floor very cheaply, or "squat" illegally without paying any rent, usually working upward, a practice called "piggybacking."

Max explains: "The first place you wanna set up is the first floor, or the basement. Then you wanna take a second floor apartment and a third or fourth floor apartment, if you can. You move to other floors 'cause the first floor takes all the heat, and when it falls, you got the other floors as backup."

Cocaine-amatic

"The place is open twenty-four seven [24 hours a day, seven days a week]," says the young man who works behind the scale for Splib's friend Victor. He is describing how discounts work. "We sell fat twennies [$20 packets] for three weeks to get customers—we give forty [$40 worth] for twenty the first weeks because we know people will tell other people. Then we stop selling twennies, we up it to thirty at least. Then we go to selling seven lines [a gram] for seventy dollars for three-four weeks, then back down to five lines. After that we stop because we don't want no more customers.

"I hope you don't think we don't protect ourselves," he gestured to a large shopping bag of money near a table filled with cocaine. "We have nine millimeters and a machete. We have two bodyguards. We don't search nobody, but we look at them real good and one bodyguard is at the front of the line and one is at the door when people come in." Asked about current prices, the boy was reluctant to commit himself, but Victor, who arrived a short time later, spoke freely about the prices as he wrote them on a piece of paper.

"My prices right now is one-half 'cuarto,' which is three and a half grams, that's one hundred fifty dollars. One 'onzo,' which is twenty-eight grams, that's a thousand dollars; one-half onzo, fourteen grams, is five hundred twenty-five. A cuarto, which is seven grams, that's two hundred seventy-five. And one-half 'octavo,' which is sixty-two and a half grams, that's one thousand, seven hundred dollars." (These were 1987 prices, dramatically down from 1986.)

The houses mark a significant change in the cocaine trade. Over the last few years, most dealing has moved from individuals acting alone to institutional arrangements like the houses and to vertically organized groups or crews. These often include members of one family, but even when that is not the case, they maintain family-like structures.

Packaging has also had its effects. Pre-packed cocaine and crack in small units is now the standard, and suppliers who want the product to reach the customer undiluted deliver these packages in tamper-proof sealed containers. Selling pre-packaged cocaine avoids arguments about weight and purity, and eliminates the need for hardware, such as scales and cutting paraphernalia at the retail level. Sealed packages also save time for both the buyer and the dealer.

Finally, street-level dealers are now usually paid all their wages in cash; suppliers trying to curtail overuse no longer pay with cocaine. This also means that dealers no longer give out samples—as one buyer commented, "Coke is like a regular business now. Dealers don't give samples, people come in and know what they want. The stuff is there prepackaged, it's good, and they know the price. They pay the money and leave."

Scramblin in the Street

"All the kids out here be scramblin," Kitty points out, lighting a cigarette. "It's a way of life. If you need some money, you start scramblin." As we talk, Kitty has been selling cocaine in the street for two years, as has Jake; Max and Chillie have worked in the trade for six years, Splib for ten. Kitty tells something about the odds against surviving on your own, without an organization.

"Nobody stays in the street for years hustling like this and makes money without a crew. If they stay out there, they are hooked in some kinda way to the life, the excitement, the nickel-and-diming, or the drugs. Right now, if you stay out here a few months you doing good.

"I don't believe most people make money over the long run, only in the short. You get high, spend a little something here and there, but you don't make the kind of money you expect to when you start. I mean on the street level, on the kilo level, on whatever level, because the more you get the more you use up. The real big guy in the crew is the only one who really makes money."

Kitty admits she's not making a lot of money, but she keeps dealing because "I like to think I can make money out here. I know I've been lucky so far, and I like to think my luck hasn't run out yet."

Being a woman presents special problems. "I deal with a lot of men in this business, and you wouldn't believe how Latin men are about women dealing. It's very hard." It is true that almost all street dealers—and runners, touts and lookouts—are male. And all men, whether Latino, African-American or white, argue that the risk and danger require aggressiveness, intelligence and cunning—virtues, except the last one, that males consider their own.

Kitty challenges all this, as she has had to do in other

settings. "My father, God rest his soul, tried his best to keep me at home and be a wife to him and a mother to my little brother after my mother left him. But I couldn't do that. My brother was his responsibility, not mine."

Jake's Troubles

Jake is working near 162nd Street. "Jumbo crack. I got that jumbo crack," he says to no one in particular, glancing furtively over his shoulder. He works the Heights, and knows the money-making potential in particular places. This summer has been a good one for Jake: he began selling 200 $5 vials of crack a day and ended up selling 500. He deals from 6:00 pm to roughly 1:00 am every day except Sunday and his customers are increasingly repeats. He lives on 156th Street, in the heart of the trading area, and only has to go downstairs to sell his wares.

When Jake scrambles in the street, standing on a corner or side street well known to buyers and police, he will oftentimes wear a distinctive piece of apparel, like a Mets baseball cap, so buyers who have been referred can identify him. Many buyers drive by and ask about price, size and quality.

There are many beat artists selling bogus drugs in these copping zones, so buyers will remain loyal to an honest dealer. To get repeat business, street dealers must quickly establish a reputation for "excellent product." Since the transaction is now too quick to allow the buyer to taste or test the goods, buyers express dissatisfaction by not returning, or by asking for a discount next time. One early sign of a dealer messing up the work is when he cheats regular customers by cutting the drug too much, by selling short, or overcharging. Buyers will often not buy again from such a dealer.

The phone in the corner booth rings and Jake reacts as

if it's a gun going off, jerking his head around before he catches himself. He is convinced that phone is monitored by the police, so he tells his customers never to call him from it and does not use it himself. He does answer it now, however; it's Max calling to tell him to come over to the apartment right away.

By the time he arrives, Chillie, Charlie, Max and Masterrap have already prepared most of the vials. Kitty has not showed up yet; Jake is toting the day's receipts in a brown paper bag. "Who's cooking?" he jokes.

"I am," Max answers with a disgusted look. "You getting ready to pack?"

"Yeah, I guess so. You got the caps?"

"Yeah, you know I do." He buys 5,000 of these little capsules at a time from a "guy who works for a company that makes them. Sometimes, I give him coke."

The table is littered with tiny plastic vials. Jake barely looks up as he counts three small off-white chips of crack, seals the top with a red cap, then flips it into a pile of several thousand others. Suzanne, who rarely gets involved in the business, is busy counting "caps" and looking over the operation because Max has to make a run downtown. After an hour of this, Jake is sweating and wipes the perspiration from his brow as he sings along with a Spanish song on the stereo.

He talks about his street experience. "When I first used to go out there, only the kids was there. If they didn't buy for themselves, they buy for somebody—you know what I say; they was kids. Now I get people from everywhere that buy. The other dealers made my thing change. One day I go out they be selling for ten dollars a cap. The next day it's eight dollars a cap, some days it's been five. Last year I would sell a few over an hour and people would mostly buy the coke. But now, word! Hey, I be selling thirty-forty caps in a few minutes."

Max tells a different story. "Jake's been having problems taking care of the work lately. Last night he came back with a hundred fifty vials. Something's wrong. I think he needs a rest; he's been doing too much. I thought he was okay until his cousin told me how he has been up all night a lot and stuff. That's not like him.

"He's been talking a lot, too, and you know he don't have that much to say. His cousin told me over the last few weeks he's had these girls and he's spending money and stuff like that. He's going to Dominica for a while and Charlie will handle some of his work."

Jake was not just a teenager "acting out." He was spending too much money and snorting too much cocaine. Dealers often overindulge—it is easy for them with large amounts of the drug readily available. Max, seeing the symptoms, moves to correct the situation as quickly as possible by cutting back on the amount of "work" Jake receives. This greatly reduces the amount of cocaine partying he does: as Max knows, dealers have an aversion to buying cocaine; most of them believe that only a fool, a "chump" will buy it. Many dealers say they find it hard to believe that people pay such prices, no matter how good it makes them feel.

Max thinks Jake is acting this way because he now relies on cocaine as a sign of personal worth. The girls, he says, are only attracted to Jake because of the money he makes. "The minute the coke stops and the money stops, those girls stop coming around," Max says with certainty.

While he recognizes Jake's need to assert himself, Max does not want any mistakes that might jeopardize the entire operation. Max, all of nineteen, is playing the role of sage. "I have seen this happen many times. My brother was only one. Every time somebody starts to make crazy money they discover the girls, they discover the big ride, they want to take control. Word.

"I have a problem myself keeping everything together, but not many people listen to you when you tell them what's right. They got to go through it by themselves. Jake is only seventeen, he's still a kid in many ways though he's been through a lot of shit.

"At that party at the spot, he pulled away from that dancer when she was trying to get him to go with her. Not just that time, but many times with the females he's been real chill and shit. He isn't no faggot or nothing; he just felt those girls wanted him because he had the coke and was my 'back.' But now I think he wants to break out."

Jake apparently does intend to break with Max, but has not decided when or how. Many breaks do occur—over women and money and drugs and loyalty—but the Kids have a resilience that transcends these problems. Their disagreements are like family quarrels and, as in a family, they eventually come together in crisis or celebration. There is a strong feeling of kinship beyond ethnic bonding in the cocaine business in general, and among the Kids in particular. (And there are real family ties: Max and his brother Hector are related to Jake by marriage; Chillie's cousin is a dealer, as is Splib's brother-in-law. Max's uncle was a dealer.)

"When I had that dream that time," Jake is saying, "I went right out and bought me a nine millimeter. Didn't I Max?" Max, annoyed, looks up and says something under his breath; clearly, he is more than a little embarrassed and somewhat puzzled by Jake's behavior. For the moment, he ignores him, partly because he must cook up more crack and partly because he doesn't want to confront whatever is hiding behind Jake's new-found self.

He may also be puzzled by Jake's lack of obedience. Not very long ago, Jake snapped to at Max's every command. Jake has had older people tell him what to do before. "I used to belong to the Ball Busters when I was thirteen,"

Jake explains proudly. "That crew was mostly around 134th Street at that time. I was with Colt [the leader] and we had some old Dominicans telling us what to do. We were a tagging crew [graffiti artists] and we would do gang-banging [fight with other crews over wall turf] and shit like that. Colt was bugged, man, he would do anything.

"When I think about all the shit—fighting and fucking over people and running from the police—it was crazy, it was just kid stuff. I remember when we went down to the Roxy before this big dance and Bambaata [leader of the Zulu Nation crew] kept calling his crew, he was really messing with us. So we pulled off our jackets and showed them our colors [each gang distinguishes itself by wearing beads of a particular color]. It was full of Zulus but the ILL Boys and the Savage Nomads all joined us. It was two hundred of us and we went outside. I snatched mucho beads that night—I still got 'em at home now."

Jake goes on to say that though he works for Max, nobody owns him. He complains that everybody wants him to be the one to go to the *grocería* to buy beer and "do all the shit work because I am the youngest. I'm tired of being treated like that.

"When I got the nine millimeter, I told everybody. I wanted them to know I was no pussy and not to fuck with me. I wasn't going to hurt nobody, but I didn't want to be soft—like they thought I was because I did what Max told me."

MAX is tired. Chillie is late paying for some cocaine, and Max needs the money for his supplier. Suzanne is still angry at Chillie and Max is wondering if he should keep Jake on the crew—he is pissed off at Jake, but refuses to confront him.

But he has to mix. He pours cocaine from a baggie onto a sheet of glass, adds baking soda and comeback. "This is an eighth of a key, and the comeback is only forty grams. Ain't nobody out here putting that little amount of comeback into their crack, but I do, and I only put in sixty grams of baking soda." He puts the mixture in boiling water in the glass pot and covers it; now he must wait. He is feeling discouraged.

"I don't know what's wrong with everybody. Yesterday, everything was fine; today there are motherfuckers everywhere. In your job, you go in, do what you have to do, and then it's over, right? Not me. I'm glad you're seeing this part, because most people think it's just sniffing and spending money all the time. This is hard work. It's hard because you've got to deal with so many different people."

3

The Kids

The Kids live in flux, with the day-by-day troubles, the sorrows and expectations, of any adolescents. Their individual stories emerge when they talk of their lives and their experiences in and around the cocaine scene.

Kitty and Splib

Kitty often talks about her life with Splib. He has his side of the story, but somehow Kitty seems more sincere and truthful, her actions less part of a "feminine" game than Splib's use of "masculine" protection and disguise; certainly he rarely admits being wrong.

Kitty may express her feelings freely because she comes to me as a last resort, distraught and exhausted, reeling from some encounter with him or after days of running in the street—trying to get away from her attachment to him, but unable to do so. She wants to lay her heart out to a sympathetic ear, but also to someone who can get word to Splib that she is suffering, to let him know she is in pain over him.

"I know what I have to do with my life. My mother, God bless her soul, would say to me, 'don't let people take advantage of you.' I think she was saying this because of the way my father treated her. He had a real mean temper and he would beat her every time he got drunk. My father was cheap, too—he would bring home these real cheap clothes and shoes for my brother and me, and my mother would ask him why he didn't spend more money. But he would get mad and beat her up, so she stopped saying anything. The only time he was really nice to her was when she was sick.

"My father is a typical Dominican man in every way except he was so fucking cheap. Dominicans like to save money and send it home to their families, but they are not cheap, especially when it comes to their children. But my father would work and spend his money drinking himself crazy with his friends.

"When I think about it now, I don't think he was a bad person. Not really. You know, when you're a kid you are looking up at everything and everybody. Everybody is bigger than you are, the talk is bigger than you, the laughter is bigger. And the sadness is bigger than you are.

"Now I know what went on between my father and my mother was between two people who loved each other, but I couldn't understand that as a little girl. All I saw was this man hurting my mommy. Now, as a woman, with what I'm going through with Splib, I have a whole different perspective on things.

"My mother did everything in the house. That's what she knew to do—if she could have done anything else, I really believe she would have. She did the housework, the cooking, the whole thing, and she took care of us kids and my father, too.

"My father was nothing when my mother left him. He couldn't wash a dish, he couldn't clean the house, he couldn't go to work—he was like a zombie. I was twelve then, my brother was four. My father wanted me to do everything and I did it.

"But that's when most of my troubles started, around twelve. I used to see my parents fucking all the time, and that really excited me. I couldn't wait to try it myself. I was fully developed early—I had big tits and my ass was sticking out and my menstrual cycle came when I was ten. My parents were very strict about boys, and I couldn't think about having boyfriends while my father was around. And that was one reason I left when I was seventeen—that and because he wanted me to spend the rest of my life cooped up taking care of him and my brother.

"My parents were very religious, spiritual. My mother was just Catholic and I don't have to say any more about that. My father believes in *Santería*; he's a *Santero*, and I sort of picked up from him about the saints and the orishes [god-like intermediaries between the Supreme Being and humans]—you've seen the prayer bowls and everything in the apartment. I don't take it as far as he did, with the chickens and goats and all the sacrifices, but I do believe there is power in it.

"I went to public school and I liked it at the beginning. We had a program where every student was free to study in or out of class. We had these sheets, one for each subject— you would have certain material to read by the end of the week, and maybe a book report each month. You could take the day sheets home or do your homework sheets all in school. It was a good way to learn, but it didn't last long—the teachers who were doing this were transferred or something. And then we moved a few blocks down the street and I couldn't go to that school again.

"After tenth grade I was fed up with school, and I missed a lot. I lied to my father about it, and after a while the lies caught up with me. He was mad as hell and tried to force me to go back to school. We kept arguing, so I left home for good. I started to hang out with my boyfriend and his friends. We would go to his parents' house while they were away at work and sniff coke and have sex and do whatever we wanted. Right after that, I learned about hustling, fast money and how to play games. That was when I met Splib."

Kitty and Splib were married in 1982 and shared an apartment in upper Washington Heights for about three years, then separated in 1985 with some bitterness. One problem was Splib's ego, his need to conquer all the women he saw, and to so dominate Kitty that she could not pursue opportunities for love or money. Kitty says she only wanted a family, and to be with him, but somewhere—amidst all the hustling and the hassles, arguing and fighting, deals and dives—they lost the love they began with.

Splib

Talking with Splib revealed yet another side of the problem. He would say Kitty was not a good mother, but she clearly was; he would say she didn't love him, but she clearly did. Kitty would often just react to Splib's treatment by going off for several days and—he claimed—freebasing or engaging in other degrading activities.

He says his son deserves more of his attention than he has been able to give. To Splib, the essence of life is the family—how you take care of your family and how you look in their eyes. "How do I take care of my family when everything is up against my back? How do I look my kid in the eye when I don't have enough time to play with him? Kitty knows I need to make money out here in order to be with him."

One reason he wanted to make as much money as possible, Splib once told me, was that his mother needed an operation and he wanted her to get a specialist. I loaned him $50, believing I would eventually be paid back. But, as I learned later, Splib's mother was not dying of some dread disease; this was one of the cruel ruses he employs to con money from friends, cocaine from dealers—and anything valuable from anybody. Splib rarely, if ever, repays the money he borrows or cons from people. Even when it comes to paying for cocaine received on consignment, he is always late.

Once, when Splib reneged on paying for an ounce, Max said, "Something is always happening to Splib. We all have times when we have troubles, but this guy, word, every day he has a problem. I swear." Max continued, with a forced smile, "His sister needs an abortion. Her husband needs cash. His brother is in jail and needs bail. He needs a job. His father is sick and needs money—that's funny, because as long as I've known Splib he has never said a word about his father; I don't even know if he's got a father. Word.

"Just the other night I saw him at Jump-Offs and he says his apartment had been robbed. He called six of us together at his sister's place to ask if we would help him out. And, you know Ramon? [Ramon, the supplier of last resort, is considered a bit crazy and none of the Kids but Splib would deal with him.] Ramon gave him five grams of rocks, and we all gave him two hundred dollars apiece."

Although Splib is known for his daily lament of disasters, he is still respected for his keen wit, intelligence and an uncanny ability to emerge unscathed from the most difficult situations: Max, Ramon, and Chillie all know of his poor record, but they come through to help him. Max says he has given up on him many times, and yet when Splib called, he came—with money. This is perhaps indicative of the closeness of the Kids, but that closeness can quickly turn to animosity when money, women and self-image are at stake.

Ramon explains, in his heavy accent, "To live in A-mer-rica, you must have money my friend. I have tried to work for the white man here, I have tried very hard. But the more you work, the richer he gets. I don't like when you try to make me a fool. Splib owes me money. It's not for me, the money; I have to pay my supplier. It don't belong to me. Why should I have to pay for his problems?"

Ramon is known as a man given to extreme acts of violence against anyone who happens to get in his way. He is especially known for shooting those dealers he has trusted with cocaine who renege on their promises. Just a week earlier, Ramon had threatened to kill Splib if he didn't return two ounces consigned to him. "I'm going to fuck him up if he don't give me all my money or the *perico*. I warn him, 'if you don't bring money you lose your life.'" Splib managed to keep Ramon pacified until he was able to pull the same trick he had pulled after his birthday once again.

At one point, perhaps because of his falling reputation, Splib took a regular job working with a Dominican automobile mechanic in the Bronx. He helped the man paint, clean, and weld cars. "It's a job," he said. "Roberto [the garage owner] deals a little blow on the side, too, because so many of the dealers give him coke as a way to pay him and thank him."

But Splib didn't last long with Roberto: not for the first time, he said he was earning too little money; this time, he claimed, Roberto himself was the problem. "The dude, I swear, would sniff coke all day long and into the night. That was alright, but shit, he wanted to keep working like to four in the morning—shit like that. Fuck it, man; all I wanted to do after I got high was to keep getting high, not work. You know, Roberto would have these girls over to suck his dick, and he'd fuck and then he'd jump up and say, 'OK. Let's paint this bad boy.' Forget about that. I just split."

And went quickly back into business. "A few days later, I picked up a little package from a guy I knew years ago, Juan Victor, a little short Dominican dude. So I got me a thang to put out in the street. Juan Victor wants to open up a coke house. I told him to go for it, I'm in."

The Birthday Party

Splib often asked me to come to his house to eat the guinea hens he said he was good at cooking; this day, there is a special occasion: the second birthday of his son, Armando.

We are in the six-room apartment Splib shared with Kitty until they separated a few weeks before. Splib's mother, sister and friends are helping cook the guinea hens. He has also invited his current girl friend, Irene, claiming that he didn't expect Kitty since he hadn't talked to her for a week or more, but of course she appears.

Kitty and Irene do not like each other, for obvious reasons, and each has accused the other of interfering in her relationship with Splib. Splib is unmoved by this for the most part; in fact, he feels women should fight over men and that he is certainly a man worth fighting for. He claims that Kitty learned all she knows about cocaine from him, and would not be making any money in the street without his support, and has often intimated that he is the strength behind the relationship. But an outsider can see that it is Kitty who has the maturity, the wit, the dominating personality—and enough love for Splib to keep his ego massaged. For the moment, things are fine; they are cordial and enjoying the party.

The room is a rainbow of colors. Splib's mother is in a mauve and blue dress, Kitty wears a red pleated shirt; colorful streamers hang from the ceiling and Armando sports a crooked black bow tie and is draped in an ill-fitting black suit. He is uncomfortable with all the attention and pulls on the tie until it comes off. His grandmother comes over to fix it; he pulls at it again. Six other children hover around him, run and jump, scream and cry; two little girls fight over a balloon. The grandmother arranges the children under the *piñata* hanging from the ceiling. The kids jump up and down trying to pull the string of ribbons suspended from it. Finally, one of them reaches it and candy, balloons, coins and dollar bills fall to the floor as they scramble for the goodies.

Kitty comes over to help her son gather some of the loot. She looks like a kid herself, playing, laughing and rolling on the floor with the children, her skirt ruffled by the scramble; her strong face with full lips is accented by a bright smile. Her hands, however, are rough and masculine, with jagged fingernails.

She and Irene barely acknowledge each other; so far, this is the only sign of tension between them. Irene and Splib talk but avoid touching. She is plainly dressed in tight-fitting cuffed pants and a blouse that is remarkably like Kitty's. Splib is in an all-white suit, with a large gold chain around his neck.

Kitty speaks to her mother for a moment, then walks over and kisses Splib, imprinting a lipstick-red outline of her mouth on his cheek, glances at Irene, and leaves.

THE party is a sort of focal point, with emotional lines of force converging on it and emanating from it.

Shortly after the party, Kitty disappeared for a few days—neither her mother nor Splib knew where she was. She called me and came to see me to announce she was leaving Splib for good, and asked me not to tell him she had been to my place. An hour later he called to ask if I had talked to Kitty. "I checked the after-hours club where she used to work," he said with a tinge of nervousness in his voice, "but they haven't seen her. You know, they turned that spot into a crack house, it's just crack and more crack. I told her at the party I didn't want her working there and she said she wasn't, and we had a little argument. It's been three days since I've seen her."

This is not the first time Kitty has disappeared, and

Splib knows it will not be the last, but he is genuinely worried his own actions may have pushed her to run away. He says he is off cocaine now because he will not snort in a depressed state of mind. He talks as if he thinks that rejecting the drug is a way to conjure her back to him. But this will last only as long as no one offers him cocaine.

"I know she prefers me to cocaine," he says, in a pained voice, "I just know it. She would come to me when she wanted to get high and we would get high together. But now she does all the blow by herself. It's harder selling blow when she's around because we both start to get high and before you know it we ain't sold nothing.

"She sniffs like crazy now. But you know I don't care about all that. I just want my kid to know she's his mother." He imagines the worst, "It's been three days—I'm afraid she's on the pipe, and that shit's bad news. I worry about the basing. She used to sniff only a little bit, but now she has no discipline. I feel that basing will completely destroy her will to say no; I believe that pipe will take her away from me."

Irene and Splib

When Armando is with Splib at his apartment, Irene pays special attention to the baby's care, and criticizes Kitty: "*Ella quiere ser una puta. Ella es floja, y no se preocupa del niño.* [She acts like a whore. She's lazy and she doesn't take care of the baby.]" She apparently thinks this will endear her to Splib, but it has no noticeable effect on him one way or the other: he says he needs the help of both women.

It is true that Kitty often leaves the baby with Splib for hours while she sniffs and parties—not an activity that will win her any good mothering awards—but she argues that she needs a break from all the hassle, and that Splib has as much responsibility for the child as she does. When Irene is there, she will change Armando every time he is dirty or wet, and spend as much time playing with and coddling him as she does with Splib. He often says that she cares more for Armando than she does for him. She has a daughter, age three, and seems to be more maternal than Kitty; they are the same age, but Irene displays patience, a quality Kitty rarely exhibits, and then only in a forced and dutiful way. Kitty is always in a hurry.

On Saturdays, Irene and her daughter arrange to spend time with Splib and Armando. Splib is often late for these appointments; sometimes he doesn't show up at all. If he is at home but busy with another girl he will not open the door, letting Irene wait for hours at a girlfriend's house down the block or at a coffee shop across the street. Eventually, when Irene calls from the street, he will ask the (usually young) girl to leave.

Relations between Irene and Kitty took a turn for the worse when Splib began receiving cocaine consignments from Kitty. Kitty had switched from Chillie to a male friend she will not name. This upsets Splib, but he liked being able to get cocaine from her because she is extremely generous and he can have his way with her. Irene is troubled because he often stays over at Kitty's, and their closeness makes Irene feel uncomfortable and insecure. She tries to compensate by buying Splib clothes and bringing him cocaine she obtains from other sources—in a vain attempt to wrest his affections from Kitty.

Kitty and Carlos

In fact, Kitty does not feel kindly toward Splib at the moment. She had asked Splib to pick up $5,000 owed to her from a buyer. "I felt that as the man he should go and ask these assholes to pay up. Why should I have to keep asking them? They won't treat him like they do me." Splib went to get the money—and dropped out of sight.

Kitty, fearing that he had been hurt by her customers, frantically called all his friends and hers, checked after-hours clubs and bars all over the Bronx and along Broadway. Then she saw her customers in the street and they assured her they had given Splib the money, and her concern for Splib's well-being turned into concern for her money: she knew neither her money nor Splib would be in the same shape when she found them.

Splib chose to reappear at their son's birthday party because she would not say much about the money among family members.

"He's always setting me up for a cross," Kitty comments sorrowfully. "I don't just mean about this little bitch, Irene, I'm talking about all the shit he does. Do you know what he told his sister about me? He said I was a whore, he said I was fucking around. The only reason he said that was because I was making plenty of money and his ass was dragging the ground.

"Nobody trusts him anymore; he couldn't get anything on consignment his reputation is so fucked up. Do you know who forgave him for all the shit he put me through? Me, that's who. I'm the one who said, 'Splib, you can have some of the money, just bring me three [$3,000] back. He brings back two. Anybody else and they would be dead, but I let it go. He knows how I feel about him, and he takes advantage of it every time. He thinks things will always be this way, but he's mistaken."

She was more than a little disturbed that Splib brought Irene to the party, but she has been depressed about all that has gone wrong with their relationship. She and her mother provide Armando's day-to-day care, and Kitty says she has given up the hectic side of dealing, and sells cocaine only occasionally. Splib is no longer giving her money or support—and when he did, the amounts were minimal. "He hasn't had the courtesy to take Armando to the movies in. . .

I don't know how long. He doesn't care if the kid's clothes are washed, he doesn't know when he gets sick with fevers. How could he? He's not here. I know I'm the mother and I should do these things; I'm just saying he doesn't realize how hard it is."

One reason for Splib's distance is that Kitty's new boyfriend, Carlos, does not like the attention Kitty gives Splib when he comes to visit Armando—in fact, Splib is not allowed in the house when Carlos is there. Carlos does not like the fact that Armando carries Splib's name and not his and wants to adopt him, but Kitty refuses, and insists on making the boy aware of who his real father is. Carlos wants Kitty to have a baby with him; she says she will in a year.

School Days

One of the few times most of the crew gather outside the apartments and away from the street is at the paddleball court in Riverside Park. One spring Sunday in 1985, I met Max, Chillie, Masterrap and Jake there to play handball, and to listen to them talk about their school days, their families and their childhood memories.

Max spent his first three years in a rural village in the Dominican Republic, where mango and banana trees are plentiful, and the kids play all day in the sun. His parents moved to New York, but soon returned to their home, leaving Max and his brother with an aunt. He is always reluctant to discuss school when Chillie is around, perhaps because Max is genuinely embarrassed about his lack of schooling and doesn't want Chillie to have the upper hand in anything.

"My moms and pops always wanted Hector and me to go to college," he says, dripping with perspiration after a hard hour on the court. "I never cared much about school because I was facing the street every day. There was a time

I'd think about school—but not that much, because everybody don't fit in school. I didn't. Chillie and Charlie like school, they do alright in there. I think Kitty would do good, too, but the one who is really smart and could do good is Splib. He's too smart to be doing as bad as he's doing. But school is out for me. I never liked the school, I never liked the teachers, and I never liked the kids."

Chillie stands against the tall concrete backboard smiling broadly as he talks with two Puerto Rican girls waiting for their turn to play. He gestures with his hands as they move toward another court. "This one will be free in a minute," he says, "and if it's not, you can come up to my house and play." They look back—one with a smile, the other with a smirk—and say, "No thanks."

"I go to school now for college credit at the City University," he informs me with a serious expression on his face. "I really wanna finish. I don't want to get trapped in this coke business. But as long as I don't do that and have my goal, I'll be alright."

But Chillie wasn't alright and perhaps he knew it even at the time. That day we all talked about drugs, sex, school, girls, life, family, among other things and he was as self-assured as ever. But several months later, he was having second thoughts about being in the cocaine trade. He said he was sniffing too much, that the females were putting too much pressure on him. As a result, school got pushed farther back. One day, he called to ask for help on a paper he had written, a short autobiography; I told him it was good, but he was convinced it was awful despite my assurances.

Charlie's feelings about school were a more complex story. Of all the teenagers I met in the street underground, Charlie was the least likely candidate for drug dealing, and most aware that he should have been in college.

One day, his day off, we talked while waiting for Chillie

to return to the apartment. Charlie wears a short-waisted bomber jacket; a white, hand-painted graffiti sweatshirt covers his muscled chest, the cuffs of his faded denim jeans rest on top of white unlaced Adidas sneakers.

"I think my mother and father would have wanted me to at least finish college. They both went to college and I know I should, too. But somewhere down the line I just didn't feel school could do much for me. I mean, I had to pay rent, I had to buy clothes, I had to live . . . and school, college just ain't it. Well, at least for now. I don't plan to be in this business forever; I've got potential to do better, and I will. But right now, the thing is to make some money. After I do that, then I can think about college and all that."

As he talked, a vivid story unfolded about his father's ambitions and efforts to hold the family together after his mother's death. His life had begun as the embodiment of the American middle-class dream, utterly unlike the immigrant background of his Dominican friends.

"My father dreamed about living in the country so we moved out of Brooklyn to a place in Long Island. Everybody said we were moving to the 'burbs, and none of my friends wanted us to go where only white people stayed. This was Long Beach. We bought a nice style house with a backyard and patio. We were the first in our family to move out of the city and we were considered the hicks."

Charlie's father was a salesman, always on the road, though sometimes, he would take his son along on short trips. "My father was a born salesman and could talk his way out of any situation. My mother was a very different person; it was like my father kept her alive by being so full of energy. You see my mother had cancer long before anybody knew anything about it. By the time my father found out she was in such pain all the time no drug could help her. I remember her shrinking in weight, and losing all the spirit she once had.

"She's been dead ten years now. I was just eight when she died. I really only knew her when she was sick. I wanted to hug her sometimes and my father would say 'don't do it' because it would hurt her. I couldn't quite understand that. My sister took it very hard because she was very close to her and not to my father as I was. She was fourteen when my mother died. I remember the day. My sister took me to the bathroom and told me. I just stayed in there a while and cried. Then I went out to play."

After his mother's death, his father hired a series of housekeepers, "but all of them quit after a few days with me," Charlie remembers. Then came Mrs. Anderson, whom Charlie and his father liked, and who was a great cook; unfortunately, his sister hated Mrs. Anderson. She finally quit after almost two years.

"My father couldn't understand what the problem was, but then he was on the road five days a week. He would get weekly reports from my teachers and from Mrs. Anderson about my bad habits, and he would sometimes whip me. My sister was always good in school. She could speak good and everything, not like me. She could do everything with math and science too. My father could never come to school events because he was always gone. My sister had lost respect for him and was always disappointed with him. Then as soon as we started to get over my mother's death, my father died. His death affected me much more than my mother's. For one thing I was older and I understood more of what it meant."

Charlie remembers the house being sold to pay people his father still owed for his mother's illness. The company his father worked for came to take the car back. That year his sister married and went to Texas with her husband, while Charlie moved in with his uncle. "He was a New York City correction officer. I liked him and he had a nice home and everything, but I just couldn't adjust to his ways, plus

I didn't like my cousin or my aunt that much. When they got divorced I went and stayed with him. One of the first things I had to get used to was the change from the 'burbs to the city. We didn't play stickball in Long Beach because we had enough room to play baseball. I had no idea what box baseball or stoopball was. The second thing was that I was polite, and I found out that doesn't make it in the city streets. So I learned fast, and what I didn't know I faked.

"We lived in Lenox Terrace on 135th Street, and one of my first jobs was delivering laundry. I would take bundles of clothes to old people in the block. I didn't do that for long because there was no money in it, and my uncle was getting crazy—he'd come home with all these different girls and then tell me I couldn't stay out with my friends; he'd be doing drugs but he'd tell me never to get involved with drugs."

Charlie was sixteen the year he met Chillie at Charles Evans Hughes High School. "The school was hard and everybody was bad. If you were white, you either didn't come to school much or kept real chill. I did all right in English and in math. I didn't like science or history much, but everything was going okay for me.

"Then one day I left some coke in my coat pocket—I didn't even do coke then. I had it because a week before Chillie gave me five [a stylized hand slap] and the coke was in his hand. We were in the hallway and it just wasn't time for conversation, plus we were leaving and so I just put it in my pocket.

"My uncle took my clothes to the cleaners and he found the stuff. He asked me did I do coke and I told him no, not really, but I knew about it. He told me okay, but never have it in the house again.

"After that, I lived in hell with him: he was always telling me I was gonna make him lose his job, and that I didn't appreciate what he was doing for me. Then one day

he comes home and says I can't go out on weekends anymore because all I did was use drugs. This was totally crazy.
I decided it was time to go. I moved out of there. I still call
him and shit but, hey, I ain't kissing his ass for a visit or
nothing. You know I can take care of myself.

"I graduated high school and got a regular job as a mail
clerk and moved in with some friends. But they were all
scrambling—I was the only one with a regular job. Every
week they'd be asking me for money to buy coke and smoke.
Then one day I saw Chillie driving on Broadway and he
asked me to give him a call. I had been taking karate lessons
and was pretty good at it, and my uncle taught me how to
use a gun—we used to go to this shooting range in New
Jersey—so when Chillie called and said he was opening up
this office and needed a bodyguard, I took it. It pays a lot
more than playing the mail clerk game."

In many ways, the Kids have become Charlie's family:
he hangs out with Masterrap away from the office; parties
with Juliette, a girl he met through Chillie; plays handball
and goes to the movies with Jake and Max now and then—
and he has come to depend on the crew for financial,
emotional and moral support.

Max and Suzanne

At about 11 one summer night in 1985, Max calls,
saying he has a personal problem and asks me to meet him
at El Cubano. It's after midnight when I arrive, passing two
kids in sneakers standing with hands in pockets, whisper-
ing "jumbo crack" to passers-by. The streets are busy, the
atmosphere frenetic: double-parked cars stretch for two
blocks, buyers and sellers are making trades, children play
sidewalk games, young mothers stroll with their babies,
women with curlers lean out of third- or fourth-floor windows to shout to neighbors across the street, men with

groceries go into "number holes" to play the *bolita*, the night numbers. The herky-jerky sounds of rap and salsa scream from the large radios teenagers carry shoulder high.

The bar stands in shadow across the street, alternately illuminated and dimmed by the nervous twitch of a blinking streetlight. Drinking and talking at the bar are Max, Chillie and Hector. Chillie waves me over, and before I can sit down, Max has ordered a round of drinks for everybody. The scowl on his face makes it clear that he is disturbed about something: the usually confident and self-assured Max looks worried and defeated.

We go to sit at a table in the back of the bar. He fiddles with his drink, then asks, "Do you know a doctor I can go to? I've been trying to have kids for two years now—and nothing." He looks up toward the ceiling and then down into his glass. "I know my wife can have kids because she's been checked by a doctor. Do you think it's all the coke I've been doing? Or do you think it's just me?"

He had said nothing to Chillie and Hector about this. To cover his anxiety, Max told Chillie he was worried because his connect hadn't delivered—an explanation that also served other purposes. Max feels Chillie is making too much of his demand for lower prices and more weight in his consignments, and this fictional problem with the connection may help cool Chillie down. If he thinks Max cannot purchase cocaine at the old price he might lay off. I try to convince Max to stop sniffing for a while (it would not be realistic to advise him to stop altogether), relax, stop worrying about his performance and see a doctor.

The next day, Max picks me up in his new Audi 5000 (I have arranged to talk with Suzanne at their apartment). One window is broken, the work of thieves who had tried to grab his radio a day or two before. "I just got the car two months ago and I've had two windows broken," he says, rolling down a good window and dumping the contents of his ashtray on the street. "You see the dash," he demonstrates. "See, the radio comes right out—it's a Bensi Box. They're stupid thieves because they don't even look in the car; if they looked they would see I don't even have a radio in here.

"You know, everybody puts signs in the window saying, 'No Radio.' Well, a friend of mine told me a joke about this mucho macho guy. This dude bought a new BMW, right, and he's so bad this guy puts a sign in his window that said in Spanish and English: 'Yes, I have a motherfucking radio and I dare you to fuck with it'." Max is in a good mood, laughing as we drive up Riverside Drive.

Suzanne is cooking paella and the smell of it greets us at the door. Most of the furniture in the apartment is new. There are television sets in several rooms, stereo equipment in the dining room and bedrooms, clock radios on side tables. This is Max's tenth apartment since I've known him, apart from the cocaine safe houses. This one has a large kitchen and big windows. Today is Sunday, Max's day to relax. He doesn't answer his phone today, only his beeper calls, and then only if he recognizes the number. Max and I sit in the living room as Suzanne works in the kitchen with her aunt.

Suzanne is tall, medium-complexioned, with a deep dimple in her left cheek. She is wearing a gold dress and high heels. Most of the gold around her neck and the diamond rings on her fingers are gifts from Max. She speaks slow, flawless English and what she calls a Castilian Spanish and is outspoken in her conversation about Max and their relationship—willing to speak freely of their sex life and his dating other girls. When she gets to the subject of other women and Max, she snaps her thumb and forefinger together and runs that Castilian Spanish across her tongue, cursing both the women and him. We had talked briefly before, but today I wanted to hear her side of their

story. I made the proper gesture, asking Max if I could talk
to her. She was excited about this chance to talk and we sat
in the dining room while Max watched the baseball game.

"My mother is Dominican and my father is Colombian," she tells me, adjusting the chairs around the dining
room table. "When I was about seven my father moved to
Florida. This was right before my mother opened a business
in New York, a small restaurant on the corner of 143rd
Street and Broadway. We used to play outside with my aunt
because my mother was always busy. When I was about
ten, she closed the place and we moved to an apartment on
Broadway at 158th Street. I loved it there. We used to go
down into the oval and shoot firecrackers, and my uncle
would take us to Riverside Park.

"My mother was very protective of me and didn't want
me to go out much. When I got to be fourteen, she said I had
to be careful not to get pregnant so I could marry as a virgin.
All my girl friends were having babies and my mother
wouldn't even let me look at a boy. She was so strict because
she could see my friends' lives weren't working out right.
You got to know what you're doing, and they were so young,
they didn't know what they were doing when they got
married or when they had babies so early. Even when I
wanted to marry Max, she said no, I wasn't ready.

"One day I want to have kids but only when Max is
ready. I don't wanna raise no kid by myself."

Sometimes Suzanne seemed a little too good to be
true; certainly she never revealed any of her own faults. But
over the years, she was loyal to Max and he never said
anything critical about her. She had every right to hang out
and party because Max spent so little time with her, but she
didn't—which intensified her loneliness.

Suzanne justified talking with me, as did all the Kids,
by saying I should get the truth for my book. I was the

sympathetic listener, the person who could hear all and everything and still be trusted. This conversation, for example, was in strict confidence. It became awkward only when Max, getting edgy about the length of our talk, heard his name mentioned, and came over to see what was being said. Suzanne held her ground and refused to repeat what had been said. When I noticed he was getting angry with her I said we were talking about his uncle—the one who was shot. He told her not to talk to me too much about that and went back to watch the game.

That evening she also revealed her anxieties about their relationship. "I know Max has to make money somehow," she offered, pausing to gather her words. "And he makes good money in this. It's hard, but you gotta take the risk. I don't complain too much because he pays the bills. It's not like I didn't know what he was doing. When I went out with him I knew he was into cocaine because I used to see him with his brother and everybody knew what his brother was into. And Max told me he wanted to make money, lots of money. Sometimes I wish he would get out of it, and sometimes he says he will, but I don't think he will. He's scared to to into legitimate business here in the United States. I don't know why. His head is like that.

"He gives me everything I want. He don't spend too much; he never goes overboard. If I were in his shoes I'd be spending like crazy. He says he's gonna help me start a business but I haven't seen any signs of it yet."

Suzanne has real anxieties about Max's future and is acutely aware of the difference between an ongoing business and flash in the pan. More than that, she suffers from loneliness, boredom, and an enforced passivity. "I just wish he would take me out more. I have to stay home until it's time for him to take me out. What I don't like is when he says he's going out for business and gets all dressed up and comes home blasted. That's not business. When I ask him why I can't go with him, and why does he lie to me, all he can say is that I'm not the type of girl to go to the places like the girls he sees out there."

Suzanne protests the double standard that permits Latin males absolute sexual freedom. "I know those girls are basically no good," she says, "but why does he go out with them? All they want to do is get high and spend his money. He says he lies sometimes so he can get what he wants. You know I caught him in bed with this girl one time. He says he wasn't doing anything. When I ask him why he does those things, he says that it's just the kind of things men do. And then he says every marriage goes through that. I don't agree with him even though it does happen. He said he gets so skied [high from cocaine], and that's why he does these things."

Suzanne's complaints echo Kitty's protest, but Suzanne tries to reason with Max because she is certainly not about to leave him. She has a somewhat defeated air, in contrast to Kitty's resolute independence. "He says he will never do that again. I told him if he did it once he will do it again. Right? But it's very boring because it's lonely being here by myself, and he's never home. Before we got married I was blinded by my feelings for him, my love for him. I didn't realize that people change as they get older and go through more things. My mother told me that everything changes after marriage. I never understood that. One thing I have learned is never trust a man because he will walk all over you."

Suzanne could easily turn to drugs for solace, but she is very wise about the habit. "I don't take cocaine or crack because once you start you can't stop. I never use any of it, I keep away from it. I've seen crazy things happen to people on it. Take the ladies—they will do anything for some crack. My uncle takes crack and he's crazy as he can be. He will stare around the house and talk to himself; he pretends he's seeing things that are not there. He's nervous and hyper all the time. I don't want to be like that."

Suzanne's distaste for crack was shared by the Kids, but her attitude toward sniffing was exceptional: almost all the young women associated with the cocaine traders used the drug regularly.

The food is ready: the table is filled with paella, fried sweet plantains, a grape sangria, rolls and wine. Suzanne's aunt comes out to say hello and asks Max in Spanish if I'm a basketball player.

After dinner, we talk a little more, and I come to understand that Suzanne spends a great deal of time at home partly because it is her "place" as a Latin woman, but also partly as a result of Max's trade. The Kids' girlfriends or wives don't hang out much, except when the boys decide to take them to discos or parties; they cannot trust strangers— indeed, they are told to be wary of everyone—so close friends, mostly other girls, become the center of their every-day world.

The community presses women to be true to their men—even imposes sanctions on those who do not remain loyal to one boyfriend at a time. Yet Max and the other boys have little loyalty to girlfriends who would limit their freedom to have other girlfriends or to participate in the sexual adventures they encounter in their daily dealings.

For the young women, there is the frustration and boredom of the endless hours, sometimes days, of loneliness—and all that is quite apart from the ever-present danger of being alone guarding large amounts of cocaine and cash, a role women are expected to perform regardless of the risk. They do remain loyal to their men, and care for the children day to day, with the help of grandmothers, aunts, neighbors, and close friends—surrogate parents, babysitters, and helpers in an extended, ever-changing family network, in a rough neighborhood of this toughest and most challenging of cities.

Masterrap and the Slang

Masterrap, puffing on a cocaine-laced cigarette, complains to Charlie, "Hey, it's canoeing"—burning only on top so the unburnt paper looks like a canoe.

"Pass it, then," Charlie throws out.

"Fuck you," Masterrap retorts, wetting his forefinger with saliva to stop the canoeing effect. He only succeeds in burning his finger and dousing the cigarette. He manages to light it again, follows the trail of smoke with his eyes, looks over at Chillie and finally passes the cigarette. His gold neck chain contrasts with his dark chin and black cap.

"My crimey here thinks the way to go is more drugs," he says; "But I know better. I think making money is okay, but not making it just by dealing. You gotta go legit, at least for a minute. You gotta go 'state of fresh,' all the way live, if you wanna do anything worthwhile out here. Everybody thinks they can make crazy dollars, but they confused. It ain't like that. I've seen co-caine bust many a head—they get fucked up and be clocking out after they find out they cannot find the key to understanding the mystery of skied. I say skied. But you know what? But-but-but you know what? They don't have a clue. Word."

[In a loose translation: "My partner here thinks the best way to make money is to deal drugs. But that's not what I've learned from my experience. If you're going to make real money, you must use illegal money to establish a legal business. You must do this now, while the time is ripe, if you really want to make it out of here. Everybody thinks they can make lots of fast money, but that's wrong. It's not the way it works. One thing I've seen on the street is that using cocaine can really make you confused—cocaine has turned people around. It has confused a lot of people and many of them are going crazy because they don't understand what too much cocaine can do to their minds and bodies. Most of them don't understand anything about the drug—they have no idea. You can take my word for it."]

Masterrap is a creative inventor of language and truly a master communicator. To him, slang is an event, like a dance step, a movement, a gestural display; a linguistic happening for other teenagers. But they, too, have to command a unique presence, not just "bite" his style. Whenever he has the chance, he performs for everyone present, though he finds an audience of young women particularly inspiring.

"My game ain't cocaine," he would say. "When I crack on a female 'how you livin?' she gots to respond to me in the positive, or I don't waste my time. Most of these sneaker bitches is looking to get skied, not looking for knowledge. I'm about knowledge. I'm about understanding. Word.

"I was talking to a female the other day with Max and all she wanted to know was, was I getting paid. And she asked me whose baby Benz that was. When I told her it was Max's she didn't wanna hear my conversation. She wanted to rap to him. I said if you gonna be hoeing [whoring] for a rap you ain't nothing but a dog bitch anyway. I didn't mean to break on her like that, but that's what she deserved. Max wasn't interested and she couldn't do nothing for me after that."

Masterrap was often playful in his word antics, but when discussing his family and his future he chose a more serious tone.

"Dominican kids have a lot of respect for their families, especially their mother. Most of the Dominicans around here come here to make money. They get into drugs—and other things, too—and then often go back to Santo Domingo, but I don't want to go back there. I been here (New York City) since I was nine years old and I like to think of myself as a New Yorker. If I go anywhere it will be to Puerto Rico. I like it over there."

Masterrap has visited Puerto Rico several times with his parents and once by himself. "When I grew up in Santo Domingo with all the banana trees and mangos and shit and all the sun all year long, it was fun, but my family was poor. We weren't poor like the poor-poor, but we didn't have that much. That place is like being in Russia somewhere. My father didn't make no money until he came here.

"I was born right up the street here at 139th Street. My mother took me back to Santo Domingo where I stayed with my grandmother most of the time and then we came stateside when I was nine. So when I save these dollar bills, I'm gonna go somewhere else.

"All the Dominicans want to go back, but not me. My father took me there to see my grandmother when I was fourteen. It was not for me; I didn't like it at all. That's small-town life—everybody know who you are, what you do and shit. You go to the same places all the time. The military control everything. They be walking around the streets with submachine guns. No man, that ain't for the kid. My mom wants to go back some day, and when she does I won't be going with her.

"Right now I want to get my rap songs out there. When Chillie asked me to be part of his crew, I told him I would do it only for two years. I said I would do it if we could make some crazy dollars. I wanted the money to make a demo [demonstration record] and go on into the record business. I've made some money now and I got my demo. After this year, I stop. It's been more than two years, I know, but I'm moving in a way I feel good about. I want you to hear my record. I got a box full of songs I ain't even took out to show nobody. I know I can do raps better than what's out there.

The Fat Boys, Run DMC, Kurtis Blow, L.L. Cool J., none of
their raps is better than my stuff. Once a producer hears my
shit, he's got to like it.

"You know, this 'paper' [money] thang ain't gonna last
forever. I wanna get hooked up with what's real. I give my
moms money from time, and I give my uncle some cash,
too, as a back. They know I work here and they sniff some
of the coke I give them, but they know how to help me out
too. They be real chill about everything, and they advise me
on things and help me when I need it.

"My uncle is like a father to me. Every since my father
passed, he been more than good to me. He was the one who
told me selling coke is just like any other business—you
gotta work hard, stay on your toes, protect what's yours, and
not fuck up with silly matters. In America you gotta have
money he says, because that's what people respect.

"My mother used to hide her coke from me until one
night I find out she's sniffing with her boyfriend. She was all
worried I would be hurt—this was long before I was doing
this with Chillie. I told her I knew all about coke, and she
wanted to know how I knew so I told her some older boys
were using it and I saw them and shit. Every since then
she's been chill. She is always telling me to be a man and
take care of my responsibilities. 'Take care of your family
first,' she tells me. I bring home money to help her with the
rent and buy her gifts."

Masterrap thinks he inherited his musical gift. "My old
man was a musician; he played guitar. My mother was a
singer, and my old man always had these musicians around
the house, playing and dancing and partying all the time.
That's the first time I even saw coke, although I didn't know
what it was. My old man would sniff with his buddies and
play music all night. My mother would sing with them and
make drinks out of all these fruits—mangos, papayas, gua-
vas, the batidos—and I used to love those drinks. And I

would put on a little hat and pretend to sing along with them. My mother brought me this little guitar, and I could swear to the stars that I was really playing.

"My mother and father broke up at least a thousand times. My mother would start an argument about some woman who kissed my father and it was on. She'd take me and some clothes and go over to my grandmother's house— which was only a few blocks away—and they'd fight over the phone. After a few days she'd be back and we'd be singing and dancing again.

"While I was at my grandmother's I never went to school. My grandmother wanted to keep me away from my mother and father anyway, so when I was there I just ate her food and we would play just like we did back in Dominica. School was another thing altogether. When I did go, my teachers were always nice to me so I took advantage of it. I was the only child at home, and the laws at school—I couldn't deal with them. I got into fights all the time with kids. I was pretty bad.

"I never finished high school because by the time I was in ninth grade I had missed so many days it was shame. I lost interest. When my mother would send me to school, I just stayed at my grandmother's house. I would miss months of school, and when they would send my mother a letter she wouldn't say anything to me."

Like other kids in the underground economy, Master-rap values work, money, and success, and is determined to keep some sense of self-worth. For him, this means standing out in the crew by forging an identity—and that is one reason he works at being the house rapper, devising words, arranging raps on paper and setting them to music.

All the Kids would rap, charm (talk to), or game to impress girlfriends; hang it up (insult) or fresh (compliment) male friends by using special words. Much of this street talk conveys their social setting: words for drugs, the police, money, sex, attitude or toughness, form a lexicon that expresses an ironic but authentic perception of their own reality. This glib verbal gamesmanship with its rhythmic cadences has been around in African-American and Latin culture for a long time. Blues and spirituals, with their use of metaphors and hard-biting double entendres, also influence today's teenage rappers.

The more creative rappers, those who maintain poise and verbal control, gain more social status among their peers. Today, new music groups offer contemporary rap variations on the old themes of love, sex, money, power, and beauty. What was once called "jiving" and heard only in pool halls, on street corners, in school yards, game rooms and juke joints has become a new musical-linguistic form: a rhythmic, accentuated rapid-fire monologue or dialogue that may or may not rhyme.

One familiar device among teenagers is to use opposite meanings: for example, "death" denotes the ultimate in vibrant attractiveness. When Splib tells Chillie he has recently met a girl who is "death when she dresses" he means she has exquisite taste in clothes and moves gracefully.

Often rap is used to win over, impress or seduce somebody. "She can rap to me" or "he's got a good rap" are terms of admiration for someone who is inventive, constantly bringing forth new ear-catching words. Conversely, "He ain't got no rap," is a way of saying the person does not have good verbal control.

4

The Scene

For the Cocaine Kids, the after-hours club is a place where street lore survives, the setting for nightly exchanges of drug-related myths. Manners and etiquette have a role in the hostile yet strangely calm world of the after-hours club. This is also where the Kids set up to expand their operations.

Jump-Offs

"We've been having trouble with McQueen at Jump-offs," Chillie says. McQueen, a woman, is the owner of the club. It is a summer evening, and I am riding along with Chillie and Masterrap; Chillie is driving very fast, both of them are high on cocaine—and I have no intention of going uptown with them in the car so I suggest we stop and talk at El Cubano.

We get to the bar at about 7. There are few patrons at this hour; the juke box plays a slow Latin tune while the barmaid looks on, bored. Once we're settled in, Chillie starts to tell me about his plan to take over Jump-Offs. "I told McQueen I was gonna go and get my crew and come back and take over the place." He speaks softly but emphatically, like a changed man, as if he suddenly realized he had to prove to himself and the crew members that he is tough. His black shirt and white tie make him look like a 1930s gangster.

"I told her we are like one, and if she takes me out there will be another one to take my place. She ain't nothing but a woman and ain't no woman gonna control me and my crew. Listen, we make all the money for her, we take all the risk, we take all the shit and we are tired of that. Those days are over, my friend. We're taking over that spot."

McQueen is shortish, in her late forties, and weighs close to three hundred pounds. She talks fast and loud and wears a saloon-type dress designed to force her already protruding breasts further outward. Though she has owned the place for three years, she now feels uneasy because her husband is no longer around to provide the muscle she needs to keep the street boys from taking over.

Chillie had been supplying McQueen with cocaine for her operation for a year and a half, and he says she owes him a great deal of money. To pay him back, she proposed that he could run the club until he recouped his money. But Chillie, for reasons of his own, wanted to make it appear that he had forced her to give him the club. Both of them were concealing aspects of the real story, but it was clear that Chillie wanted Max and the others to see him as a big-time hustler. Masterrap and Charlie had agreed to back Chillie, but they did not really expect him to go through with the plan. When he decided to directly challenge McQueen for total control, everyone was taken by surprise.

"I didn't really believe he would do what he did," Charlie said later. "I thought he was gonna scare her or something, because she's alone. Chillie don't know that much about running no spot. He can sell the coke and shit but he can't run no spot. I'm with him, but I feel any day now that lady's gonna come down here and try to fuck us up."

Masterrap agreed, and also feared reprisals. "He challenged this woman and she challenged him back. Now he's all bent out of shape because she's a woman. He thinks he can keep this place because of what he got on her. But listen, she can drop a dime [call the police] as quick as anybody and he's gone."

Max is probably the most vociferously distressed about Chillie's action, mainly because he might lose Chillie as his crew chief. As crack has become the trade's most profitable product, Max has come to depend more on Chillie and Jake. He says Chillie has been selling four ounces a week in cocaine and crack ($16,000 a week retail then), and Max's entire operation would obviously suffer if Chillie were to take over the club, find another supplier and begin to work for himself.

Jump-Offs is an unlikely place for such maneuverings. It's a dingy little dive with twin doors, one red, one blue, resting off 7th Avenue near 119th Street. There is a steady stream of customers at all times of the day and night: men and women, boys and girls, black and brown and white, students from Columbia, transit workers, all come to the red door to buy crack and cocaine. The blue door is the club: at about 3 in the morning the red door shuts and the blue door opens to a small coterie of cocaine snorters.

"Chillie's talking about his crew," McQueen says angrily. She wears a large diamond ring on her right index finger and taps the table for emphasis with a pencil held in her left hand. She is genuinely delighted to have someone listen to her. "His fucking crew. He don't know who he's fucking with. Ever since Pee Wee [her husband] died, motherfuckers been trying to take advantage of me. But I got a suprise for them. A real surprise. I'm gonna whip his

ass so bad that he will either go all the way crazy or come
back well.

"You know he must of really felt his nuts today because
you know what he tells me? He say, 'there are a coupla
things I wanna straighten out with you cause we gonna take
this joint over.' I laughed at him. I didn't want him to get me
upset. Because I know they ain't shit, no way. So, I say,
'Who? You? You and the crew!' He said if I keep on talking,
he and his crew was gonna put me outta here. I say, 'you
silly little boy. Do you know that I ain't scared of you? My
grandaddy died fighting white folk. My daddy is in the
mountains at Dannemora [prison] doing two life sentences
for killing motherfuckers fourteen times your size, and he
taught me. Now, this ain't 158th Street, little nigger, this is
119th. This is Harlem. You don't own nothing here. Ain't
nothing here belong to you.'

"He just say 'Yeah, well, I'm telling you now, bitch, we
taking over'. I say, 'You ain't taking nothing. You can't do
nothing; you really don't know what's real. You playing
cowboys or something? One way or the other you gonna
grow up,' I told him. I say, 'You lucky one day and the next
day your luck run out.' I say, 'You think everybody scared of
you. But they ain't. You can't take nothing; you can only
dish it out—well this time it ain't gonna be so easy.' I told
him he was a baby and somebody was gonna teach him how
to walk and talk one day and that teacher just might be me."

Even with the sincerity in her voice, I felt McQueen
was not telling the whole story, that she had, in fact, made
some kind of a deal with Chillie. True, her club was illegal,
but she still could have called the police if Chillie threatened
her with physical harm. Both she and Chillie seemed to be
concealing some key element of the situation. Most likely,
she had offered Chillie a share in the club to repay him, but
Chillie decided to play a little game, with two goals: showing

the world how tough he was and threatening Max with the loss of his business.

The After-Hours Clubs

In the early morning hours, before the city has washed her face, people stream out of after-hours clubs like Jump-Offs along Seventh Avenue, walking past many sprawling buildings, empty and solemn. The boulevard is lined with one-dimensional retail stores all begging for change.

The after-hours club is a community institution, a haven for all kinds of illegal and hidden activities. Gambling, snorting, basing and sex all make their way into the clubs. It must be a secret place to keep the squares out and let the hip, fresh and chill crowds in. The after-hours club has special significance not only for its regulars and its owner but for the police.

New York City has a large number of such clubs in addition to "cocaine bars" like El Cubano. They open as early as 11 pm and as late as 4 am, and stay open until the last customers leave. There is an entrance fee of $3 to $5, and drinks cost as much as $5—the clubs have no liquor licenses, but liquor, like cocaine, flows freely. Security guards screen patrons for weapons and admit only those who are known to them or have been recommended.

The after-hours club is a key social institution of the cocaine culture; users from all backgrounds frequent them at one time or another to share cocaine, engage in cocaine rituals and socialize especially with dealers and other regular users. Although dealers sometimes go to the clubs to sell drugs, most act as if they are there for recreation and take time out from selling cocaine to mix with cocaine-using friends and to share stories with other dealers.

Generally, cocaine dealers come to after-hours clubs with "C-C," or calling card cocaine, to give out. This en-

hances their status, helps them make new friends and most
of all, attracts new clients. If a patron asks to buy, dealers
typically reserve their higher-quality cocaine for first-time
sales. In many clubs, dealers—regulars themselves—de-
plete their own supply of cocaine in the course of the
evening and will often purchase "house cocaine" from in-
house dealers like Kitty.

Most customers at Jump-Offs are African-Americans,
but in recent years Latin-Americans have made their way
into the community and many live nearby. Most of these
newcomers are Dominicans, although there are some
Puerto Ricans and Colombians. In many ways, they try to
emulate African-Americans, and Jump-Offs is one of the
first places where they can mingle socially with their neighbors. As a result, there is a mixture of languages spoken in
the club, and the juke box plays Latin and soul numbers
with equal frequency.

Another group of regular visitors are the "rich folk from
downtown"—teenage and older whites, an adventurous lot
who show up every Saturday night; always at least five of
them, never more than ten. The club's employees call them
"weekend junkies."

Both the regulars and the weekenders seek the club's
special ambience, its curious mixture of excitement, glitter
and apprehension. The teenage habitué is looking for something out of the ordinary—the direct stimulation of cocaine
itself, or the atmosphere of illicit pleasure. Part of that
pleasure comes from the ease of human contact that cocaine encourages. Cocaine "loosens the tongue," lowers
inhibitions and draws the user into a setting where intense,
open communication is not only possible but required. One
expects, and is expected, to be sociable: strangers are
obliged to become friends.

Anyone who enters the club gets a "once over" by the
patrons. A young woman who is alone can expect to be

approached by young men and teenage boys who offer cocaine, make sexual advances, or simply try to engage her in conversation; a young woman sitting and snapping her fingers to the music is signaling that she would like to dance.

Most patrons have snorted some cocaine before coming to the club and are in a state of heightened excitability, ready to participate in a group scene ruled by a special code of behavior: they may be required to engage in conversation with anyone, to be gamed on, to listen to arguments, to drink alcohol, use illegal drugs and become involved in other taboo acts whether they want to or not.

The after-hours club, then, caters to those with an interest in risk-taking, a penchant for danger. It is the home of a specific code of behavior all its own; manners and etiquette are taken seriously.

The Hierarchy

The bell rings and the guard lets a couple enter the club. The two are dressed identically in white fedoras, white fur coats, white leather pants, silk shirts, and white boots. Both sport diamond pinky rings; he also wears a larger, heart-shaped diamond ring on his middle finger. All eyes are on them, all conversation comes to a halt. Only the jukebox seems unaware of their entrance. The two bejeweled newcomers sit at a table by themselves and order drinks, then life in the club resumes its normal pattern.

About twenty minutes later the bell rings again, and the door guard admits a young teenager wearing a black turban, black pants and a cape. He holds a small box wrapped in cloth. Again, a hush falls over the crowd. The tall man in white rises, opens the box to reveal finely crushed crystals of cocaine, and offers it to everyone present. "Check this shit out," Chillie says softly. It took some effort to keep cool in face of this display: Chillie is not alone; Splib and Max are obviously impressed.

Anyone who is allowed to enter after-hours clubs has a certain amount of status for that reason alone. Once inside, there is a competition, a clamoring for attention, which can include lavish spending, elegant attire, and generous sharing of cocaine. Those who can create these effects most effectively win the respect and envy of everyone present, if only for an evening.

One cold night in December I met Chillie at Jump-Offs. He was in one of his better moods, saying he was breaking from Max and going on his own. Just after he entered the club, he went upstairs followed by about five kids, not all friends of his; I followed a few minutes later and found eight or nine teenagers seated in a semicircle around several tables, talking in Spanish with a smattering of English. Chillie is telling a story about Splib.

"So Splib is sitting with his woman in this restaurant right? And as he eats the *asopao* [a soup] he notices a hair in it. You know Splib, super macho man, he calls the waitress over and he says, 'Miss, what is this?' pointing to the hair in the soup. He's getting real loud now. 'I pay my money for food and there's a hair in it,' blah blah blah. The place is full of people and it's a small spot. So it's embarrassing to the waitress, who is this big titty Cuban with her hair all curled up. She is about to leave with the bowl of soup and he says something else to her. So, she turns and says, in a loud voice, 'One hair. You complain about one fucking hair. Every night you put your face in a whole head of hair and eat it and you worry about one hair.' Everybody in the joint broke. It was funny as shit."

Chillie brings out a large foil packet and proceeds to pass it around. The other patrons in the room look on with studied interest as he gives his entourage several sniffs of cocaine. The atmosphere is one of unrestrained gaiety.

As I discovered later, it is rare for anyone to offer cocaine to more than three or four people at a time. That night, Chillie was king of the big time.

Regulars at the club work hard to maintain their dignity and enhance their status. A key element in this effort is the attempt to exude chill or cool—poise under pressure. Conversation is smooth and controlled; people temper their displays of emotion, avoid uncouth behavior, and above all cultivate an attitude of detachment. Even Chillie's story is designed to get attention at the expense of Splib.

A teenager can get immediate status by entering the club and giving large amounts of cocaine to everyone present and win even greater status by repeating the performance. As Splib says, "You know, Chillie wants to be big shit all the time now. He wants to give away an ounce of coke, have all the females on his arms, telling him what a great man he is. He wants to leave the barmaid a fifty dollar tip."

Once a teenager has established a reputation as a big spender, he is expected to continue or else suffer a loss of face or status. To avoid this, many choose to stay away from the club until they can replenish their supply of money and cocaine.

Some loss of status can also occur when a teenager breaches the rule of cocaine sharing. Chillie showed his contempt for two kids in the club one night. The two were within earshot, and he obviously intended them to hear him. "They've been in here all night," he said. "One doesn't drink, and the other has snorted up all Jake's coke—but Jake don't know it because he's so zonked out he don't know his head from a hole. One of them had a little bit of coke in a piece of foil and gave some to me. So me, like a fool, I gave them a little package of what I had, expecting them to buy a little something. Well, they both sniffed and sniffed. I was expecting them to give me more of theirs, you know, but I guess they didn't have any more. Or else they didn't want to bring it out. Can you check that out?"

There is a hierarchy in the after-hours clubs. It is not rigid—in fact, it can vary from day to day—and it emphasizes street knowledge rather than formal education. However, the real measure of status in the club is money, just as it is outside.

There are two kinds of money: one is the kind that is available in a fairly steady way and allows its owner to be generous virtually every time he appears in the club; the other is the instant wealth obtained by hitting the number, winning at dice or cards, and the like. Each conveys high status, but only the teenager who has established a pattern of heavy spending suffers a serious loss if he falters in his display of wealth and generosity; it is understood that those with instant wealth may never be seen in the club again.

Among the teenage patrons, the dealer—with a more or less steady supply of money—has the highest status. The big-time teenage dealers who spend most of their nights at Jump-Offs are both the backbone of the club's economy and the apex of its social structure.

Not far behind the dealers are the players—pimps, gamblers and other hustlers who are able to spend money and share cocaine more or less consistently. Like the dealers, they display considerable ingenuity, elegance and personal style. Another less flamboyant group of habitués are those who make money through wit, skill, and guile: boosters (professional shoplifters), dippers (pickpockets) and con artists. Then there are "popcorn pimps," a term of contempt for men who force women into prostitution, unlike true players who claim to operate with personal magnetism alone.

Generally lower in status at Jump-Offs are the youngsters who enter the business as runners between the age of eight and fourteen. These younger patrons are sensitive to the older habitués' description of them as "young turks" or "kids in felony shoes [sneakers]." Some of the older runners, age fourteen or so, open their own clubs because they feel more comfortable with members of their own age group and, more importantly, are attracted to the enormous profits such clubs generate.

At the bottom of Jump-Offs social ladder are the teens and others who make money through legitimate work— variously described as "lames," "squares" and "punks." They come to the club infrequently and rarely, if ever, share their cocaine with more than one or two others.

Staying on Top

Staying on top in the cocaine culture is not easy. Even for the most fortunate ones, high status is ephemeral. This fact of life in the cocaine business as a whole—as well as in the after-hours club, its social core—has a distinct impact on friendships.

In general, dealers maintain steady friendships with other dealers, players with other players, and so on. But a sudden change in the status of one party can damage such relationships. This possibility is most pressing for the dealers, perhaps because they have the most to lose. Thus dealers fear arrest, but their deepest concern is loss of status, not the possibility of serving time in jail. A dealer who is unable to avoid arrest loses not only his clientele and his main source of income, but his place in his clique and the glory that comes with being the supplier of a much-desired commodity.

Chillie's fifteen-year-old cousin, known in the street as "The Kid" was an aggressive street dealer constantly trying to build up his clientele so he could accumulate enough money to move up in the dealing hierarchy. One evening, the car he was driving was stopped by police, and though he and the other teens in the car threw their cocaine on the floor to avoid being charged with possession, he was arrested. The Kid lost $1,000 worth of cocaine and had to pay his lawyer about $2,000. Charges against him were eventually dismissed, but in the interim some of his customers heard about the arrest and he lost most of his clients to other dealers; those who stayed with him were afraid to come to his apartment.

"You see," Chillie explained, "a lot of them customers won't come back. They're afraid his phone is tapped or he's being watched. In a way I don't blame them. But he's in a position where he's hurting real bad now. He was making five, six hundred dollars in a few hours before that thang happened. Now an ounce he would of moved in a day takes him five days. He's looking for a new apartment because that will ease some of the customers' minds."

An arrested dealer loses contact with the community but if he is resourceful and maintains good relations with his supplier it is not impossible to re-enter the cocaine-selling hierarchy. In the interim, he can find some solace among friends and acquaintances, or with his family.

For the teen dealer, staying on top requires handling often unstable personal relationships, dealing with a constant stream of people of different ages, personalities, gender and ethnicity. There is little privacy: as long as he is selling cocaine, he must contend with the phone ringing at any hour of the day or night. A constant stream of visitors, sooner or later, makes the operation visible to the authorities, and it is only a matter of time before the dealer has to "chill out" or work on a street corner or in a bar. And once a bar becomes known as a cocaine bar and begins to attract too much attention, the dealer may be asked to move on.

After contending with all these pressures, he is expected to saunter nonchalantly into Jump-Offs and dis-

pense large quantities of free cocaine. There are easier ways
to make a living.

IN New York, and other big cities, teenagers from the
housing projects who have dropped out of school move
quickly into working with cocaine crews, just as older hustlers move from the street to the after-hours clubs and bars.
When word gets out on the street that a particular place is
dangerous, that only the toughest can "hang" there, it becomes the "in" place.

The street corner and the bar are permanent features
of the cocaine community, the boundaries of life for each
generation of dealers and often the first place an urban kid
makes a dollar. In a sense the clubs are extensions of street
territory, designed to defend against outsiders—more often
than not, the police—but clubs are also institutions in
a counter-community at odds with the rules and values of
the larger society, the moral opposite of the church, for
example.

Every community has its good places and bad places,
places for work and places for play, places for sin and places
for atonement, places for hiding and places for showing off.
For the Cocaine Kids, the after-hours club is all of these
things. That is why you will find them there, flaunting their
gold, spending money, snorting cocaine and dealing. For
Kitty, it can be a business location; for Max and the boys, an
after-work hangout. At one time or another, they can all be
found at Jump-Offs.

The Crack House and Base House Trade

Splib sits on the floor snorting, using the corner of a
matchbook cover. Between snorts, he talks slyly about the
crack houses. "You know the corner one at '17th [117th
Street]? They had bars on the door and no heat but they

were making plenty money. The place was funky and they had two guys at the door." He passes the cocaine to Charlie, who examines it, then refuses it, saying he has to work.

"I've seen that joint," Charlie says. "But the place—I know you must have seen it, it's on 146th Street, near Amsterdam Avenue. Everybody is standing in line and the guy behind the scale has a gun in his belt. He's got a scale on the table, with some cut foil, some 'caine, and a bag full of money. And in the other room they got vials."

"That little corner store with the old man out front all the time?" Splib answers. "Yeah, I know the spot. My man gots this small table with drawers next to him, and his partner is sitting across from him with a Uzi. The drawers are for the cash and to make change, and you know what the Uzi is for."

The Kids rarely went to base houses, save Kitty, who hung out occasionally at Jason's place, or Splib, who seemed to find customers everywhere. But to understand the impact of crack dealing and the intense interest in freebasing, I had to explore the houses. Splib insisted I go to one place where he knew the "manager"—no one claims ownership of these dens—assuring me I would have no trouble getting in and promising to join me.

At the time we have set, as I expected, he does not appear, so I go in alone. In the well-lit hallway, it is possible to see blood stains along the edges of the dirty marble floor—they are fresh; not yet caked. On the second floor landing, a sharp odor oozes out to greet me. The uninitiated could dismiss this as some foreign, unidentifiable pungency, but I know now it is the rough chemical smell of "base." The door opens a crack before I can knock, a tall African-American man brusquely thrusts his palm toward me and asks, "You got three dollars?" He motions excitedly, "If you ain't got three dollars you can't come in here." The entrance fee. I pay and walk in.

The establishment is desolate, uninviting, dank and smoky. The carpet in the first room is shit-brown and heavily stained, pockmarked by so many smoke burns that it looks like an abstract design. In the dim light, all the people on the scene seem to be in repose, almost inanimate, for a moment.

As my eyes adjust to the smoke, several bodies emerge. I see jaws moving, hear voices barking hoarsely into walkie-talkies—something about money; their talk is jagged, nasal and female. One woman takes out an aluminum foil packet, snorts some of its contents, passes it to her partner then disappears into another room. In a corner near the window, a shadowy figure moans. One woman sits with her skirt over her head, while a bobbing head writhes underneath her. In an adjacent alcove, I see another couple copulating. Somewhere in the corridor a man and woman argue loudly in Spanish. Staccato rap music sneaks over the grunts and hollers.

The smell is a nauseating mix of semen, crack, sweat, other human body odors, funk and filth. Two men dicker about who took the last "hit" (puff); two others are on their hands and knees looking for crack particles they claim they have lost in the carpet.

In the crack houses, the sharing rituals associated with snorting are being supplanted by more individualistic, detached arrangements where people come together for erotic stimulation, sexual activity, and cocaine smoking. They may be total strangers, seeking only brief and super-ficial physical contact, encounters designed to heighten sensations; the smoking act is a narcissistic fix—there is little thought for the other person. The emotional content is largely due to the momentary excitation of the setting and the cocaine. Much of the sexual behavior is performed to acquire more cocaine.

Nothing better exemplifies the new attitude than the

act of *Sancocho* (a word meaning to cut up in little pieces and stew). To sancocho is to steal crack, drugs or money from a friend or other person who is not alert, a regular practice in the crack houses. Another example is the "hit kiss" ritual: after inhaling deeply, basers literally "kiss"—put their lips together and exhale the smoke into each other's mouths. This not only saves all the valuable smoke, but also stimulates the other sexually. Other versions of the kiss extend to other orifices.

Kitty describes a crack house, one of two places owned by Jason. "It was in this apartment in the Bronx, on the Concourse, a big place with a sunken living room. It had five rooms for people to sit around and get high. At first people came after they left the first house, and most of them was already zooted up [high]. They would pick up girls from his other place and come to this place to base. Most of them, I would say, the majority just come to fuck because Jason advertised it like that. There was almost no regular joints, like after-hours clubs, that were base houses; they were almost all converted apartments.

"Jason charged $50 just to come in. It was full of young girls—fourteen, fifteen, sixteen year olds. Some of these girls stayed for days at a time, getting high and having sex with these guys. But most of them would fuck just to keep getting high: when they ran outta money to cop, they would go to the first man who would give them another hit and do whatever he asked—blow jobs, suck on other girls, you name it.

"They would be begging Jason to give them more coke or money. And the Spanish men! Jesus. You know what Spanish men like? They like to see two girls together. So they would tell the girls to go with the women they brought with them. After about a month or two, the place was full of just baseheads who would stay there all day or night, day after day, spending their money. There was nothing but baseheads wanting another hit."

Basers—alone or in groups—appear to talk or interact much less than do snorters, and what they say is oftentimes unclear and less focused. The basing houses function like the shooting galleries of heroin users. They are open 24 hours a day, allowing consumers to continue taking the drug without interruption over long periods of time. Consuming more base, in turn, greatly increases the likelihood of dependency and many patrons do appear to be seriously dependent. Splib's girlfriend Irene describes a scene she witnessed outside a base house.

"I was coming from downtown and saw light-skinned Jerry the gambler from Yonkers standing across the street. He used to gamble all the time when Snake had his spot. Chicago and Bigman and a few other people were standing there with him, near the crack house. I could see these cars with two white cops in them, but only Jerry saw the cops creeping up; so he started to move as fast as he could away from the steps of the joint because he knew what was about to jump off. But before he could get out of the vicinity, they grabbed him and took him back up in there. I don't know why they didn't take Chicago and Bigman.

"Well, they got six rooms in that crack house and JoJo the one who's got no teeth, she got a small room in the back, and she charging five dollars for anybody to use her part. Chicago say it be nasty-filthy in there because people piss anywhere and garbage is piled up in the corners. The funniest dude there is Dippy because he be fixin the stuff for Jerry and the others who gamble near the spot on 116th Street. Dippy cooks it up for you and charges five dollars plus a hit. People say sometimes he'd rather have the hit than the money.

"In about twenty minutes the police came back out with bunches of folk, about five at a time. I never seen so many people come outta such a little spot. It was nine o'clock in the morning. The last one I saw come out was Fat Willie. I hadn't seen him in a long time and he wasn't big, he was obese. But he's so thin now it's pitiful.

"Chicago said the cops come to this house every other day. You know what happened after the police brought them all down? They all waited in front of the crack house. Somebody musta asked the police if it was alright to go back up in. The cop just shrugged his shoulder. So, in a few minutes they all went back in search of the white cloud."

"Crack," says Kitty, "has caused the whole thing to come tumbling down. It's brought all the heat on everybody. The police, the dealers, the sniffers, everybody. Before people started that basing shit, everybody sniffed and that was that. People was sensible; people was polite. But now, kids killing their mothers and everybody to get something for the pipe. It's crazy. It's affected everything—the coke houses don't get as many customers—everybody is scared to go up in there because of all the guns and shit. They protecting themselves from the stick-up kids. Nobody wants to take the last risk, I mean to die and shit, for no cocaine.

"It's true the baseheads have made a lot of money for a lot of people, I know. All I'm saying is that the party is over because of them. We can't make the kind of money we used to make out here—I know I can't, because I ain't gonna take that kind of risk. And I know a lot of other dealers doing the same thing. If dealers could change things, it would be back to the sniffing days when the customer wasn't afraid and the dealers took them in and enjoyed a sniff without any problems."

In addition to using pipes, cocaine smokers take the salt-like granules of base and roll it with marijuana into joints that are smoked like cigarettes; others prefer a *bazuca*, in which the drug is sprinkled on cigarettes or joints and smoked. New methods, new rituals, and a new argot have emerged around these practices including exotic ways of regulating the duration of the "smoke." The regulars at one South Bronx base house are called "balloon heads" because they take the excess smoke from the pipe and blow it into a balloon, holding the balloon closed with a finger until they are ready to inhale again, using the balloon as a way to temporarily store the precious smoke.

All indications are that smoking is here to stay. The "balloon heads" may be a sort of idiosyncratic aberration, but freebasing involves considerable ritual: cooking the cocaine, drying and cooling the base, care and cleaning of the pipe, etc. Since the average freebaser is more interested in being high than in making preparations, they will seek less ritual, not more.

There is evidence that crack users, many of them teenage girls, are moving from sniffing, smoking and injecting cocaine to snorting heroin. The most popular form of non-injected drug use, smoking crack, also quite frequently leads female users to unprotected sexual behavior with intravenous drug users, as crack reduces inhibitions while creating a desire for more drugs, and male users often barter drugs for sex. This puts these women at increased risk of acquiring sexually-transmitted diseases, including HIV infection. In short, crack use can indirectly put teenage girls at risk of acquiring AIDS.

Although crack had become popular, the "in" drug, by 1985-86, as the Cocaine Kids reached their late teens, there was still a viable cocaine-snorting culture when Kitty split off from Max and the crew to begin her own operation.

Kitty and Dial-a-Gram

Dial-a-Gram allowed Kitty to establish her independence, gave her a way to make some real money, and bolster her own ego instead of Splib's and Max's. Her frustration with them came to a head when she went to Max to repay what Splib owed for a cocaine consignment. Splib begged her in his own inimitable style ("Bitch, why don't you listen to me?") not to go see Max. But he could not dissuade her from physically handling the money to Max. Splib wanted to pay Max himself to avoid the embarrassment of appearing soft, not quite manly—and to let him make it appear as if the money was his. Kitty insisted that the main point was to pay the debt so Splib could start working with Max again, and she was not about to trust Splib with thousands of dollars again.

Kitty understood how fragile her position was: men controlled the cocaine business and she had little power to do anything unless she went totally on her own. Dial-a-Gram was her chance; ironically, it was a man involved in prostitution who helped her gain her independence.

"I met this guy Paul one night at Jump-Offs," she explains. "He happened to like the coke I gave him to taste. He told me I was in the wrong end of the business and needed to move up to a higher-class clientele. I've heard this rap a thousand times from hustlers, players, assholes of every type; you name them. Anyway, this one was no small-time player, he had a limousine service, he said, and knew girls who 'worked in the hotel business.' It didn't take long to figure that out—you know, hotels and limos.

"But I knew he wasn't lying about the limo service because when he asked me for some more coke, and I told him I had to go and get it, he said his man would drive me. I told him no thanks, but I did see a limo outside the spot and it was the same name as the one on the card he gave me.

"Two months later, Splib and I were having problems. Splib kept telling me to get out, so I say to myself if this bastard tells me to get out one more time, I'm gonna get out. Split. So, sure to hell, he tells me I wasn't shit, that I was fucking around, and this, that and the other. And up to that time I wasn't doing shit except taking care of business in the street, coming home and doing everything I was supposed to. But while he sits and waits for me, he gets paranoid about what I'm doing." She lights another cigarette and blows the smoke toward the floor.

"My baby was staying most of the time with my mother, we only saw him on weekends. Out of the blue I call this guy and he sent the limo for me. At first, I felt so depressed and out of place because this big car pulls up—and in my neighborhood, everybody looking. Then I felt powerful, in a funny way, getting into this thing and everybody looking, even though some of them were pretending not to look. I saw that *puta* [whore] Sonia that Splib had been fucking looking out the window. I told her in my own way to eat her heart out.

"I didn't see Paul until about two hours. I just sat in the car and waited with the driver. When he did arrive we went for some drinks at a restaurant in the Village. Finally, I could hear what he wanted: he wanted me to be desperate, in a desperate situation, because this is what he uses to get the girls to do what he wants. But he wasn't a player in the way we know them. He was what he called a 'a protector'; he would offer the girls protection from any harm when they went on 'the assignment.' I asked what the assignment was and he said it was another way of saying 'on a date.'

"I didn't understand all of what he was into, but I asked a lot of questions. He said it was not illegal but involved some risk. He knew girls who are called by an agency to go out on dates with men—the men are very rich, and many of them are famous. He intercepts these calls and sends his girls out; when the agency girls arrive, they just go back and say the client didn't want them. I told him I wanted no part of prostitution business, and he said it didn't involve sex all the time; the clients often only wanted somebody to talk to or go out with. I still told him no; all I wanted to do was sell coke, and that was it.

"Finally, he said he had connections but that I would have to give him a percentage on every sale. I told him I had my own connect and could keep on like I was already doing. At this point, I could see he was getting upset with me, so I asked him to tell me what else was involved.

"He told me he had a Colombian connection, and a book of the girls' clients who used coke. He said they didn't mind if the coke was cut and there was good money to be made. He would hook me up with the girls whose clients wanted coke, and I could go with the girls on calls, make the sales, and leave. After each sale I would give him four hundred dollars on every thousand I made."

Kitty decided to try this arrangement. "He gave me a beeper and a limo to use. I would go to some places that you would not believe. I have never seen so many mink coats, bodyguards, marble floors, in my life. I always saw these things on TV, but not like this.

"Anyway, for a year I been collecting my own list of clients, and I still go with the girls and sell the coke. Most of the girls do coke, too, but they don't buy it when I take it to the clients.

"There is a time limit on when somebody's drugs have to be delivered; you can't take all day. If you're more than an hour late, the people really don't have to buy the stuff. Most people be high when they call and usually they're down to their last hit; then they call and expect me to be there before the drop comes [before the drug wears off].

"This one guy was gay—he did hair and was some big-time stylist. A nice man. He always called about two in the morning. He would always ask how soon I could be there, and no matter how fast I said I could come, he would say 'come sooner and I'll give you a nice tip.' I would call Jeff, the driver, and he would zip me down there.

"I always had to be dressed really good, so I had my beeper, my high heels and I would go. Time is everything in this business. One time I lost some coke because I was in such a hurry. I went to this guy's place, he was a client of Francine's who usually wanted two grams. He wasn't from New York I don't think because he was always at the Plaza. He always wanted me to bring pure, no flake, just rocks, and he paid one hundred fifty dollars a gram plus cab fare. I jumped into a cab that night because Paul had the cars tied up. In the elevator I always check my stash to make sure it's in my hand so if I run into an undercover scene I can throw it away from my body. This time, I check my bag—no coke. I freaked.

"I went back down, and walked out of the hotel, hoping the taxi might still be waiting, but it was gone of course. So I went across the street near the park, called Francine and told her what happened. She said the guy didn't want two anymore, he wanted ten because he had two other girls coming over and was going to have a party. I couldn't go to Max because he closes shop at 3 am—Suzanne won't let nobody see him or call him after that, especially on a Saturday night. so I went uptown to 163rd Street and bought half an ounce [14 grams]. I only had one hundred twenty dollars to my name, but Juan is a friend of Max's so he let me have it. I get back to the Plaza and the guy gave me two thousand dollars, and asked me to stay awhile.

"I've been to the Plaza, the Helmsley Palace, the St. Regis, the Doral. All kinda places. I went to the airport hotel near Kennedy one night with Joann, one of the girls I know real good. She said it was an hour call and I didn't have to do anything but wait in the car because we were going out later. When we get there, she goes upstairs and in a few minutes she's back telling me there are two guys and they want to buy some coke. I go back up with her and find out these guys are Argentinians. They start to speak Spanish; then they ask me if we know anybody who has cocaine. I told them I happened to have some and they sniffed a little and then started to laugh. They just wanted to see if we did any coke.

"Then the tall one goes into the back room and I say to myself 'this is it. I'm gonna die.' But before anything, he's back with a leather bag in his hands. He opens the bag and guess what was in it? Yeah, pure flake. We only stayed for two hours but they gave us $2,000 each. They said they would be in New York for three days and asked us to come back later. I didn't go back but Joann went with Matilda and they say the guys gave them a coupla ounces of coke apiece and $1,000 each.

"I never told Paul about the tips and stuff I made, but I was afraid to keep the real coke money from him because one of the girls might betray me. You know how vicious women are. Really. I'm still doing okay with this business."

5

The Kids
Move On

The Shooting of Chillie

I called Chillie to wish him a happy birthday and learned that the police had been to his house the day before; they looked around but did not take anything or arrest him. He said it was a bad omen. I went to see him.

"The *federales* [US Drug Enforcement Administration, DEA, officers] came last night. I answered the door; Charlie and Masterrap were at the movies so I was here by myself." The place looks as it usually does but Chillie is edgy, like a person who has had too much cocaine or smoked too little crack, or just needs a drink.

"You know what I just did? I spilled salt on the floor. Do you know what that means? It means [he spoke in Spanish] that you will not have any food to eat. And coming after these *federales* came here, it was just a bad sign, you know. It's just bad. I opened the door for them because I had nothing to hide. They knocked; they asked me if I was selling drugs."

He knew their visit meant he would have to return to the street, and he dreaded it. He paused for a moment with a far-away look in his eyes. "They came in and looked around and said my neighbors had complained about people coming in and out; that somebody on this floor had called the police saying somebody had a gun in the hallway.

"I told them I don't know nothing about this, but they said they wanted to look around. So I let them. Could I have said no? If I say no, they think I'm hiding something, right? I've got to move now. I've got to chill out a bit, too. I still go to school, but I ain't doing to well."

Chillie did not move, but less than three months after we talked he did begin selling on the street. Not only did he not like this, he felt it was beneath him, a man who once had workers dealing for him, a payroll, a reputation, respect and pride. He said he planned to find another apartment to deal from.

Federal surveillance was only one of his problems. Gradually, he began to talk of other things that were pressing him into corners. School was the first casualty—initially the morning classes, then evening; finally he wouldn't go at all. Right before the police visited, Kitty had taken several ounces for a deal and had not returned when promised (she did eventually bring his money). His clients were avoiding him now that street word was out about the police. And he was sniffing more than ever.

For several weeks he had dreams about Charlie's dog, the one that ate the cocaine—Chillie, angry, took the dog to the park and shot it—but they stopped. "Thank God, I don't have those any more." All these events, especially going back to the street, had the force of prophecy for Chillie, as they would for anyone. But he was also superstitious: he not only talked of spilling salt, but of rabbits' feet, Saint Christopher medals, and *Santería* symbolism, and he started to wear *elekis,* spiritualist colored beads, around his neck.

About two months after Chillie had been forced onto the street, Masterrap called to announce calmly, "They shot Chillie last night." The two of them had been working on 145th Street, Masterrap said:

"We had been selling good about three hours when this kid comes up and says he wanted some weight. So, I said chill, and I called Chillie over and he said okay. We had been standing by the phone booths, you know the ones they taped in the voice saying this phone is monitored by the New York Police Department. We were standing there and the kid said he was waiting for his partner with the money. We waited a few minutes and his friend shows up.

"Chillie asked them what they wanted and they said two ounces. He said they would have to wait down on the street because he didn't wanna take them up to the apartment. But the big kid with a cap on said he wanted to make sure the package was chill, otherwise he wouldn't put his money in. Chillie didn't want to blow the cash, so he said okay—at first he said no, but that paper is powerful, so he okayed it. The other thing was that we had been downstairs a long time so it was a break for us.

"So, we go upstairs and as soon as I open the door, Chillie asks me if I tied the dog up. It caught me off guard, you know, because I knew and he knew the dog was dead, and why is he saying this. Before I catch what he meant, I said 'what dog?' and by then they knew it wasn't no dog and we was just fucking with their heads. We walk in and Chillie, instead of going to the bathroom, goes to the kitchen still trying to play off this dog thang. He goes in, closes the door—the whole thing—and I think for a minute he had them. What Chillie does is he hides the piece in the bathroom, and he usually goes in and gets the piece, you know, and he walks out with it and shit; that cools any moves anybody's got, out. But he don't do this, because he wants them to think there's a dog there—I don't know. Anyway we get to the living room and the big kid says where is the coke and Chillie says where is the money? And the big kid reaches in his coat, pulls out a little piece, like a twenty-two.

"He tells Chillie to get the shit, and points the piece at

him. So Chillie stalls and they go through the place looking. The skinny kid is saying in Spanish how he didn't wanna touch the closet where all the *Santería* stuff is. The big kid says fuck it, and rips out a big bag and the coke falls out on the floor. They go to pick up the bag and Chillie tries to snatch the arm of the big kid with the gun and the kid pulls away and shoots him right in the side. They grabbed the stuff and split.

"Chillie is bleeding and holding his side. We go outside and try to get a cab. There was more blood than I ever knew was in somebody—my shoes, my shirt, my hands were all bloody and Chillie looked like he fell in a big tank with blood in it. The first cab wouldn't take us, then another guy from the neighborhood saw us and took us to the hospital."

When I went to the hospital a week later two attendants demanded identification including my driver's license and passport, and by the time I got to the room Chillie was asleep. Once he left the hospital he did not resume dealing because he was not physically able to maintain the hectic pace. He had a punctured kidney and was exhausted most of the time. He went to the Dominican Republic to spend time with his family. A few months later, during a day out swimming, he accidentally drowned.

Splib

By 1986, Splib was ostracized more than ever. He was notoriously late in paying for consignments, when he paid for them at all, and he never returned phone calls from his creditors. He drank too much and snorted too much, his money was always short. He maintained a peripheral attachment to the crew, getting some cocaine from each member, and still exaggerated about his street prowess, skills and charm.

Finally, he left cocaine dealing because he knew he

could no longer make money at it once he lost his discipline and realized he could not control his own use. He caught himself saying one day that he was only trying to make his money back, and he knew this was an admission of failure, that his craving was not for money and all the other things he professed to want, but for the cocaine.

In the spring of 1986 he called to say he was getting married in June to a young woman I'd met earlier. He asked me to the wedding, but he also wanted me to ask Kitty, who had custody of their son, to let Armando go to the ceremony. The night before the wedding he called from a hotel, high and wanting to talk, and told me he was having second thoughts. On his wedding day, his fiancée called to ask me to get him at the hotel, because the time for the ceremony was drawing closer, "and if he don't show up my family is gonna kill him." Splib made it on time, but I could not convince Kitty to allow Armando to go.

Since the marriage, Splib has worked as a part-time short order cook and tried his hand at running his own construction company. At last report he was working as a house painter and plasterer; he and his wife now live in New Jersey with a baby born in 1988.

Hector

In the latter part of 1986, after returning from another trip to the Dominican Republic for therapeutic care, Hector began basing again. He developed a novel approach to handling his craving—which he attributed to the fact that he had no money, could not find a job, and lacked will power. He went to Max's friends to get cocaine, saying that Max would repay them. His consumption had already reached levels the Kids could not bear to witness; one small hit would send him reeling on a binge sometimes lasting for days.

Yet Hector was an engaging young man who often told exceptionally funny stories as he described his own failings and weaknesses: he would read about cocaine, he said, and find every passage gave him the urge to get high; the long conversations about cocaine when people were waiting for the drug to appear had the same effect. Even seeing base pipes in a store window would create a craving.

Max's friends began to call to tell him his brother was creating heavy debt, and asking Max for "call money." Some asked as much as $5,000. Max confronted Hector, and told him they were no longer brothers, that he never wanted to see him again. He had done this before, but this time he felt Hector had become nothing short of a junkie and wanted him out of the house. After this, the mention of Hector's name was sure to bring a scowl from Max. Still, Hector stayed with his brother until his problems with freebase surfaced yet again. He eventually went back to the Dominican Republic to live with his parents.

Max

Max accumulated a great deal of money during his five years of dealing, and sent most of it to his family in the Dominican Republic. Though he was reluctant to establish a business in New York because he feared federal authorities would find him out, he finally opened a business; the realization that he could make money within the law, without risk, helped convinced him to leave the cocaine trade.

He was also moved by Suzanne's pleading with him to stop dealing, especially after Chillie was shot. The long hours without Max, the constant phone calls day and night, the frantic nights alone thinking about the police breaking down the door, the waiting and wondering if he would ever come home again, had taken their toll. In many ways, their life had become unbearable in spite of the money and the drugs and the good times. It was no longer fun, she claimed, and, stubbornly but with growing certainty, Max began to see it, too.

Six months after Chillie was shot, Max decided to leave the cocaine business. It was not an easy decision. He worried about pressure from the "old man," his connection, who had reaped huge profits (Max estimated about $8 million) from the efforts of Max and the Kids. Max felt he had to approach the matter of his resignation with caution. He once told me of a rumor about how Colombians never let people retire—once you start with them, nothing short of death or imprisonment allows you to stop.

His worries proved unfounded, however. The old man, tiring himself, felt a bond with Max and far from discouraging his departure, gave him cash so he could relocate (Max never revealed how much), and wished him Godspeed in his new life.

At one point, the old man had even asked Max to join him in a car rental business, but Max declined, as he had several other offers—perhaps because he did not want people to know he could barely read and write English.

Max also felt the trade itself was getting too dangerous and complicated. A successful cocaine business these days meant diversifying, he pointed out. You had to get into real estate—to control the extremely profitable "house" operations (bringing in up to $1,000 an hour around the clock in 1984-85), dealers actually bought apartment buildings outright. Strong competition meant employing more and more backup; the police were no longer "loyal to the dollar" or to the crews they extorted money from.

Crack trade, in short, had become a definite turnoff, and the market was changing from young white buyers from New Jersey to an increasingly sleazy crowd who begged for a hit and needed what had become a cocaine fix—an element the Kids and Max were not accustomed to and did not like contending with. These factors and Suzanne's warning that he would probably be killed if he did not leave the trade convinced Max to move to Florida, where he and Suzanne bought a house. They still have no children.

Jake

Of all the crew members, Jake seemed most likely to establish his own cocaine business: he was the youngest and had been particularly close to Max, who taught him a great deal about the business. In the Kids' heyday, he had been naive and at times unsure of himself, and at one point posed grave problems for Max when he began to overindulge, spend too much money, and lose business.

It was rumored that once he learned to trade on the street, he would telephone Max to say the crack was selling at one price (say $8 a cap) and Max would instruct him to sell at $5; Jake would sell at $8 and make $3 extra on each capsule. In time, Jake recruited a crew of new arrivals, mainly young Dominicans, who were hungry and wanted to make money. He also bought into established crack houses and set others up with his own crew, and continues to distribute cocaine and crack in the South Bronx.

Charlie

After Chillie was shot, Charlie tried for several months to stay on and run *la oficina* with Masterrap, but it proved too difficult, especially as the police visited them regularly. Competition for customers had grown intense, and many of the young women who came through started cocaine tabs that quickly reached hundreds of dollars. Masterrap and Charlie were not good managers; they could not keep up the volume of business they needed, nor could they overcome the desire to party. Eventually, Max stopped delivering cocaine because they did not keep the books properly and did not keep the cash flowing as Chillie had done. Charlie is now attending a community college in New York City.

Masterrap

Masterrap, though his talent for composing rap songs made him a favorite with "the females," did not make it into the record business. He now lives with his girlfriend in Washington Heights and works as an assistant to a chef, a Dominican who often bought cocaine in El Cubano and convinced Masterrap to get out of the cocaine business. The chef promised to help him learn, and to take him on as an assistant in a downtown restaurant. Masterrap's girlfriend was pregnant and he needed to make steady money; taking a regular job seemed the logical thing to do. He says he makes $600 a week.

WITH the loss of Chillie, the breakup of the group was a foregone conclusion because he was in his own way as important as Max in keeping the Kids together. Chillie was there for everyday happenings, while Max controlled things more from afar; Chillie was the leader among the troops, Max was the general. Chillie kept Charlie, Masterrap, even Kitty and Splib—for a while anyway—loyal to the operation. It was Chillie who helped the Kids deal with a certain degree of power and verve, and from the moment he lay wounded, a spiral of defeats began. The Kids' dream of making it out of the Heights and into fame and money became more a dream. They all knew this, and believed it was inevitable— except Jake. For Kitty, separation from Splib was another element in the demise of the group: she decided it was time to go on her own.

Kitty's New Life

Kitty talked about the end of Dial-A-Gram.

"I was making two to four thousand dollars a week then. And Paul would come over here and never wanted me to weigh the package. Once he got here, he would take some of the material out and say I owed him so much. Then he would come back a week later and say I owed him for the whole package. He was so slick, that fucking *maricón*.

"Jackie was here one day when he said I owe this and that, and she say to him, 'let me see those books.' She looked at the books and said, 'wait a minute you owe her money.' The guy was pissed off. He said, 'hey, I'm not doing business with you, I'm doing business with her.

"When he used to bring those keys I could have ripped him off. But nooo, I was Miss Honesty. I was stupid. If I knew what I know now, I would have taken the shit and run. I made just as much money with tips when I was doing Dial-a-Gram as I did with the cocaine. Not all the time, but some of the time I would make $300 to $500 in tips from those guys. I would take them coke and they would say, 'no don't go,' 'stay for a while,' 'let's talk' and let's do this and let's do that. I remember this one guy all the girls call Wildman Dave, he would tell one of the girls he wanted ten or twelve grams, and they would beep me. So, I would go there. He had a beautiful apartment on the top floor in midtown, a doorman, one of those drop down type front rooms, and I would take the ten or whatever and he would give me $1,000 cash. If I brought twelve he would give me $1,200 plus car fare. I would cut the shit before and make an easy $800 on the deal. It was great."

Kitty had to learn about the subtleties of dealing to an upper-class clientele. "Most of the clients of these girls didn't care very much about the quality of the coke—they just wanted to have it because they knew the girls like cocaine. The girls would tell me those men would try to use coke to get them hot and have sex.

"I would make a couple of hundred a day with them sometimes, all these older men—in their forties—you know, they all wanted cocaine whether they sniffed or not. Like these Japanese businessmen. All the girls loved them because they always had lots of cash.

"But after I started to make like a thousand a night, the girls got a little jealous. Francine and Joanna, for example, wanted me to give them a percentage. Then after a while all of them wanted a piece. Same shit, different day. Then Paul got busted and I was on my own and all the girls started to do their own coke business thing—they could see it was big money and they stopped calling me."

Kitty admits it was more than just competition from the girls that made her quit Dial-a-Gram. It was the night life, the constant pressure of going into strange apartments never knowing what was going to happen. And it was not being with her son.

On occasion, Splib went with her, waiting in restaurants and bars, but he wasn't dependable. At one point, before she moved in with Carlos, he did try to reconcile their differences, Kitty told me.

"Splib told me he wanted to get back together, but I think it's not possible. I tried to overlook the past and I cannot do it—I mean, fuck the past, look at the present. He's humiliated me too much." Puffing a cigarette, looking more and more perturbed, she continued. "I told him that I didn't want to share him with anybody—I mean especially physically, but he has refused to accept me, my feelings.

"You know, Splib has never treated me like he wanted

me—he always wanted something from me. If he had his way, he wouldn't work at all, he would just have me do everything, including make all the money.

"He's always worried about my relationships with women. Do I like them? Do I have sex with them? And I told him a thousand times, but he still does not believe me. He always finds some excuse to accuse me of going with some man or some woman. This coke thing is something, and with Splib it's fucking him all up sometimes.

"We broke up twenty times and every time I gave in, every time I forgave him. You know why? I figured a warm dick to come home to was better than being out there in the cold street looking for something you didn't really want. I wanted him, and only him. So I compromised, as they say—I gave him money, I gave him love, I gave him a baby. I gave him everything. But I wasn't getting nothing in return.

"When I first met Splib, I was in pretty bad shape. He brought me out of that. I had been through adventures, money, races, tests of endurance. He found me, helped me and calmed me down. He made me feel like a woman. He made me disciplined, he made my tears go away. Oh, he supported my fucked-up behavior, too, and spoiled me.

"When I wanted to get married, we did. I would come home exhausted and the house would be clean, he would massage me and make love to me so good. But when I got enough nerve, I would say to him, 'when are you gonna get a job?' He got a job dealing blow with Max, but he fucked that up. Then he got a job with Cesar for a time. He would work with the material, but there was always a problem, because every day he needed money for this or that. I said fuck it; I might as well be selling the shit myself, I told him. But he would make love to me again and I would give in again.

"No. We are not ready to come back together. For

what? Why? So we can fight and kick and scream at each
other for another five years?

"When I look at the baby, it just breaks me up because
that's all I really wanted was to be with him and our baby.
But that seems impossible. So I hope he lives his life and we
can be friends but I won't share my bedroom with him
anymore."

Kitty and Carlos have moved to the Bronx with their
baby daughter and Armando, who is now five years old.

"The kid is real smart now, you know," says Kitty, "Do
you know what that little poppy said to me the other day?
'Mommy I not gonna tell my teacher about this'—he puts
his hand to his nose, like this, snort snort—'I'm not gonna
do like that girl did to her mommy.' You remember the story
about the little girl who turned in her mother and father to
the police, somewhere in Texas I think it was. He saw the
story on TV—the mother and the father had done some coke
or heroin and the little girl saw them. Armando says, 'I don't
want you to go to jail.' "

Kitty is settled for the moment, though she says she
feels vulnerable and uneasy in this new relationship. She
knows Splib is gone from her life forever, and she is trying
to please her new man, to make the relationship work. She
does not want him to feel jealous about her past relationships, and she makes it clear from the way she addresses
him and her housewifely manner that she wants to please
him. Her new boyfriend is her security for the future, she
says. "I'm getting older now; I can't afford to do the things
I did before. I don't want to do the things I did before."

Afterword

This book set out to answer some specific questions about what kinds of kids sell drugs and why, how they get into the cocaine business and stay in it—and how they get out of it—and to give some idea of the rewards for those who make it up the ladder.

The answers we have found are not only complex but incomplete, as the lives of these young people are still unfolding. It is clear that the Kids who left the business were those who had a stake in something: Max had a wife, a house and money; Masterrap a job and a steady girlfriend; Splib a wife and a new baby; Kitty a new husband and a new baby; Charlie is working for a college degree—even Hector left in the hope of finding a new life.

All the Kids except Jake have also begun to live outside the underground; for them, I believe, the cocaine trade was only a stepping-stone to the realities of surviving in the larger world.

Have I romanticized these kids, made them more "human" so the world will understand them and the problem they represent? That has certainly not been my intention. I have tried to present the reality behind the newspaper and television version of teenagers selling cocaine on New York City streets. Yes, there is violence and death on those streets, but there are also struggling young people trying to make a place for themselves in a world few care to understand and many wish would go away. I have tried to present an honest picture of what I saw and what the Kids revealed to me over five years.

The Cocaine Kids, and many of the kids coming behind them, are drawn to the underground economy because of the opportunities that exist there. The underground offers status and prestige—rewards they are unlikely to attain in the regular economy—and is the only real economy for many. Certainly, they have no illusions about the "quick" dollar. They know the work is hard and dangerous; there is no such thing as a quick dollar.

Any solution to "the drug problem" will be difficult, but any solution that does not consider the real world of the kids must fail. This book has tried to present a calm and uninflated look at that world, without using the language of problem-finding and without offering ready-made answers (like "build more prisons," "develop a pharmacological block" or "arrest all drug users"). Unless we begin to see the life as it is, to understand how the drug trade affects, and is affected by, children (and the adults around them) we will not be able to look for solutions in an open and creative way.

TODAY, as I walk through this city of fallen dreams and unquenchable hope, to the neighborhood where I first met Max and the Kids, I see a new generation of Cocaine Kids in faded jeans and unlaced sneakers, draped with gold chains, their arrow-pointed haircuts topping fresh faces and hard-edged frowns. These kids are grown before their time, wise before they leave home, smart before they go to school, worshippers before attending church, rule-breakers before they know the rules and law breakers after they know the law.

On the corner of 162nd Street, three boys and two girls

shout to me almost in unison, their outstretched hands revealing their wares, "got that coke, got that crack, got red caps, got blues, got yellow ones—you choose. What you want, my friend? What you need?"

The innocence of the young is lost in Washington Heights these days as a new generation of street corner boys and girls enters the shadowy world of dealing and prostitution. A new generation of Cocaine Kids is embarking on a voyage, searching for dreams that most will never find.

Glossary

baby Benz small Mercedes Benz automobile, usually model 190E

baggie bags plastic bags used in packaging cocaine and other drugs; also called "Seal-a-Meals"

base cocaine with the hydrochloride removed

base crazies a kind of hallucination that leads an individual to search for the smallest particle of cocaine or crack in the mistaken belief that they have lost some of the residue; also called "picking"

basing smoking cocaine base; "dry" (or "wet") basing refer to absence (or presence) of liquid in the pipe

bazuca a cigarette rolled or sprinkled with cocaine or base; also "bazooka"

beamin getting high on cocaine

beat artist person selling bogus drugs

beiging chemically altering the cocaine so that it looks brownish, for those who think this indicates purity

bite copy another person's style

blow cocaine or coke

blunt a cigar leaf filled with marijuana

borrowing to steal

breaking to take something to excess

buda a high-grade marijuana joint filled with crack

bum rush to stampede, or crash an event ("the Brooklyn posse bum rushed the party"), or mob a person or object

canoeing describes a cigarette or joint (usually cocaine-laced) burning too rapidly along one side so it looks like a canoe *chalking* chemically altering the color of cocaine so it looks white, for those who think this indicates purity

chill very cool; chillin means relaxing, to chill out is to relax, be calm

click to come through at the right time for a friend (or a date)

clockin out acting crazy; mentally "gone"

coke house a place where cocaine is sold openly

comeback a chemical used to adulterate freebase

connect (connection) a high-level supplier of drugs

coño damn!

cop to buy or purchase, especially drugs; also a policeman

copman drug dealer who has fallen from grace but still has access to high-level suppliers—but can only get drugs for cash, not on consignment

crack cocaine base (with the hydrochloride removed) or freebase; also to say something

crack gallery a place where crack is both bought and used

crack house a place where crack is used

crack spot a place where crack is sold but not used

crash strike a person

crimey a friend who engages in illegal acts; a buddy whether engaging in illegal acts or not

def exquisite; the ultimate

doggin performing sex acts for money (usually said of a teenage girl)

drop a dime to inform (call the police)

D-T detective

'due residue; the oils trapped in a pipe after smoking base

five-o the police

flake tiny particles of powder cocaine; distinguished from rocks

flame cooking smoking cocaine base by putting the pipe over a stove flame

freebase base; freebasing is to smoke base cocaine

fresh   to look good; to compliment someone

fronting   taking drugs on consignment

hit   to inhale smoke, as from a freebase pipes; also one snort of cocaine

hoe(ing)   whore(ing)

illin   suffering severe mental stress

in a minute   time of no certain duration; in the future—a day, a week, an hour, but not in a minute

jammin   a party; also a battle

jumbos   large vials of crack sold on the streets

key(s)   kilogram(s)

lines   a rough unit of measure used by dealers (a fraction of a gram); also the way cocaine is prepared for snorting

map   strike a person

maricon   (roughly) motherfucker (in Spanish)

mula   a chemical used to adulterate base

p.c.   part commission on a drug sale

pebbles   tiny rocks in a cocaine package

perico   cocaine

player   a kind of pimp; a very poised and sophisticated person; one who controls his life without much apparent effort, has money, a car, women.

posse   an organized group of teens; a term used by Jamaican marijuana crews

recompress   to change the shape of cocaine flakes so they resemble "rock"

reconstitute   a complicated method of chemically changing cocaine so it resembles "rock"

rock   cocaine in crystalline form; pure cocaine

roll up   the sudden appearance of a group of kids on a scene

Scotty   cocaine or crack: "I'm going to see Scotty" means "I'm going to get some cocaine"; from Mr. Scott of Star Trek

scramblin   a form of hustling on the street

shake   adulterated cocaine

skied   intoxicated (high as the sky) particularly from over-indulgence in snorting or smoking cocaine

smokin   elegantly dressed

snatched up   apprehended by police

spot   an after-hours club

steerer   a worker who directs ("steers") buyers to a place where drugs are sold

throw down   to fight or have a physical confrontation

torch cooking   smoking cocaine base by using a propane or butane torch as a source of flame

tout   a worker who purchases drugs for buyers

tripping   to get high; also, to suffer from delusions of grandeur

twenty-four-seven (24-7)   open at all hours every day

vict   victim, often used as a verb: "We victed this guy"

washing   chemically altering the color and taste of cocaine

wollie   cigar leaf filled with crack or base

word   a reinforcing word, a word of agreement; similar to "amen"

zooted up   high from cocaine

Note: a number of terms relate to sexual behavior, including:

buzzin a nut	sexing me down	bonein
getting busy	eatin	hittin it
pushin up	gimme some leg	do the bozack
zeek freak	get off the bra	
vamp	(or jock) strap	

Index

After-hours club, 93–116
AIDS, 25, 111
American Indian Museum, 25
Arrest, 103, 104, 117
Astroni's, 23

Balloon heads, 111
Base house trade, 105–111
Beat artists, 40, 135
Bicarbonate of soda (baking
    soda), 17, 60
Black cats, 22
Boosters, 102
Bootleggers, 25
Boricua College, 25

Cabarrojeno, 22
Calling card cocaine, 39, 97–98
Call money, 14, 122
Canoeing, 86, 135
Catholicism, 65
Charles Evans Hughes High
    School, 78
Chill out, 104
Church of the Intercession, 25
Cocaine
    bars, 97
    calling card, 39, 97–98
    houses, 51, 52, 105–111, 135
    manufacturing of, 16–17, 40
    psychopharmacology of, 14
    recipe for, 17
    recompressed, 40–41, 136
    suppliers, 19, 21, 32, 34,
        36–39

Colombians, 19, 22, 32, 114,
    123
Comeback, 17, 61, 135
Cop men, 33, 135
Copping zones, 14, 25, 56
Corn whiskey, 25
Crack houses, 51, 52, 136

Death, 91
Dextrose, 25
Dial-a-Gram, 111–116, 126–27
Dippers, 102
Dominican Avenue, 22
Dominican Republic, 74–120,
    121
Drug Enforcement Administra-
    tion (DEA), 117

Education, 26, 65–66, 74–79
El Cubano, 97, 125
Entrapment, 46
Esperanza (Hope) Center,
    25–26
Ethnic bonding, 59

Fort Washington, 23
Freebasing, 33, 39, 47, 97, 136
Fronting, 34

Gambling, 97
Geographic Society, 25
Guards, 33

Head shops, 24–25
Hispanic Society, 25

"Hit kiss" ritual, 108
HIV, 111
Hustling, 24

Jail, 103
Jiving, 91
Jumbo crack, 79

Lactose, 25, 42
Landlord-tenant disputes, 26
Lesbians, 27

McKim and White building, 25
Mannitol, 25
Marijuana, 25, 110
Marriage, 84
Monarch Bar, 23

Numismatic Society, 25

Oficina, 16, 19, 27–30, 124–25

Pan American Club, 22
Paraphernalia, 24
Perran's, 25
Peruvian flake, 15–16
Pickpockets, 102
Piggybacking, 53
Pimps, 102
Police, 43, 46, 103–104, 105,
    119, 122
Pregnancy, teenage, 25

Prostitution, 102, 113
Puerto Rico, 88

Recompressed cocaine, 40–41,
    136
Religion, 65, 105, 132
Rocks, 40–41, 136
Rules, 15
Runners, 33, 49

Safe houses, 81
Sancocho, 108
Santería, 65, 118, 120
Sex, 64–65, 84, 91, 97, 107
    bartering of, for cocaine, 107,
    111
    "hit kiss" ritual, and the, 108
Sexually-transmitted diseases,
    25, 111
Shake, 35
Shoplifters, 102
Society of Arts and Letters, 25
Steerers, 33
Suppliers, 21, 34, 36–39
    Colombian, 19, 32, 34

Touts, 33
Trinity Cemetery, 23
Trust, 16

Washington, George, 23